US Army's Effectiveness in Reconstruction According to the Guiding Principles of Stabilization

Diane E. Chido

US Army's Effectiveness in Reconstruction According to the Guiding Principles of Stabilization

palgrave
macmillan

Diane E. Chido
DC Analytics
Erie, PA, USA

Funded by U.S. Army Peace Keeping and Stability Operations Institute (PKSOI)
Submitted by DC Analytics on 31 July 2020

ISBN 978-3-030-60004-4   ISBN 978-3-030-60005-1   (eBook)
https://doi.org/10.1007/978-3-030-60005-1

Cover illustration: © Melisa Hasan

This Palgrave Macmillan imprint is published by the registered company Springer Nature Switzerland AG
The registered company address is: Gewerbestrasse 11, 6330 Cham, Switzerland

FOREWORD

"It is not enough to win a war; it is more important to organize the peace."
—Aristotle

As Aristotle accurately pointed out, lasting peace depends on what happens after the fighting ends. The U.S. Army refers to post-conflict activities as stabilization; these activities—establishing a safe and secure environment, creating a viable government, building economic opportunities, and implementing rule of law—are essential to creating an enduring peace.

As former Chief of Stabilization at the U.S. Army Peacekeeping and Stability Operations Institute (PKSOI) I took a particular interest in Diane Chido's examination of the U.S. Army's actions in post-Civil War Reconstruction for several reasons. For one, it takes a long-term historical view on stabilization activities, which is quite different from the current lessons derived from ongoing conflict zones such as Afghanistan, Syria, Libya, and Yemen. Secondly, Ms. Chido examines post-Civil War Reconstruction through the lens of the *Guiding Principles for Stabilization and Reconstruction*, the only interagency handbook on the topic, which was produced in 2009 in a joint collaboration between the U.S. Institute of Peace (USIP) and PKSOI. Thirdly, and perhaps most interestingly, post-Civil War Reconstruction is unique in that it was conducted by Americans within the United States.

Ms. Chido's research draws out many of the enduring stabilization themes, foremost being the need for political will. It is unfortunate that political will is often lacking when needed the most—during post-conflict periods when, as Aristotle emphasized, it is "important to organize the peace." However, war fatigue and the desire to cash in on the peace dividend often outweigh the political desire to commit to long-term stabilization. If political will is lacking for stabilization on U.S. soil for the benefit of U.S. citizens, how can we expect to muster it for far flung places like Afghanistan, Iraq, and Syria? This is an important question for political leadership to consider *before* the start of hostilities.

Another common, and often overlooked, aspect of stabilization is the power of women to influence post-conflict society. Depending on the intensity and duration of the conflict, the number of military aged males may be significantly reduced in the population. As a result, the heads of a large proportion of families and households may be women. This certainly was the case in the South following the Civil War, and many white Southern women went to great lengths to perpetuate the story of Southern men bravely fighting unjust Northern aggression. The actions of Southern women not only undermined stabilization at the time but perpetuated popular myths that live on to this day in some groups.

Finally, the importance of building an inclusive society for the benefit of all cannot be overlooked. While the U.S. Army in the South made a valiant effort to help the newly freed slaves, it was woefully under-resourced for the task It lacked the requisite skills in areas as wide ranging as legal reform, governance, and fiscal policy. These skills and many more were not—and still are not—organic to the U.S. Army and needed to be furnished from other areas of government or the civilian sector during stabilization, which did not happen to a sufficient degree. Furthermore, efforts that bore initial fruit, such as the Freedman's Bureau, were scuttled before they were given a chance to make a real difference.

Failing to enfranchise the freed slaves during post-Civil War Reconstruction has implications that are felt today. Similar lessons can be deduced from disenfranchised minority groups such as the Rohingya, the Tuaregs, and the Kurds. The long-term consequence of failing to enfranchise minorities during stabilization is inevitably more unrest and continued alienation.

Diane Chido's examination of post-Civil War Reconstruction through a modern day stabilization lens brings to light many of the enduring themes, challenges, and pitfalls of stabilization. Her text is an illuminating read for long-term practitioners of stabilization as well as people new to the topic. Like most U.S. Army officers, I am quite familiar with the battles of the Civil War, but I was less knowledgeable about the Reconstruction that followed. This monograph fills this knowledge gap and proves especially helpful in understanding the roots of current unrest among African Americans and the need to muster the political will to finish what we started over 150 years ago.

Brendan J. Arcuri
Former Chief of Stabilization Operations
U.S. Army Peace Keeping and Stability Operations,
Carlisle Barracks
Carlisle, PA, USA

CONTENTS

LIST OF FIGURES

CHAPTER 1

Stabilization and Reconstruction

Abstract The post-Civil War Reconstruction period is the most all-encompassing case of post-conflict stabilization in U.S. history. President Abraham Lincoln had wisely begun planning for stabilization in 1863, developing mechanisms for reconciliation and infrastructure development projects to modernize the economy of the South, such as railroad and other infrastructure expansion to blunt Southern dependence on cotton once slavery was abolished. John Wilkes Booth's bullet brought much of that thoughtful preparation to a halt. Although not defined as "stabilization" at the time, historians suggest that the Army had also learned applicable lessons during the occupation of some Southern states while hostilities were still ongoing and even prior to the war in the recently acquired territories of California and New Mexico.

Much of this introduction is reprised from the author's April 6, 2018 post on the U.S. Army War College's "War Room" blog, available from https://war room.armywarcollege.edu/articles/everything-old-is-new-again-stabilization-les sons-from-reconstruction/. Accessed on July 14, 2020.

© The Author(s), under exclusive license to
Springer Nature Switzerland AG 2021
D. E. Chido, *US Army's Effectiveness in Reconstruction
According to the Guiding Principles of Stabilization*,
https://doi.org/10.1007/978-3-030-60005-1_1

1

Keywords Stabilization · Post-conflict operations · *Guiding Principles of Reconstruction and Stabilization* · Post-Civil War reconstruction · Consolidating battlefield gains

INTRODUCTION

Picture the scene: widespread famine, millions displaced, towns destroyed, transportation and sanitation infrastructure ruined, agricultural land and property abandoned, livestock rotting, violent insurgents and dangerous brigands roaming the landscape, education and healthcare nonexistent, and corruption flourishing. Where is this? When is this? Is it Ethiopia in the 1980s? Sudan in the 1990s? Is it Northern Syria today? It is all of these, but this scene was also the reality across much of the American South in the summer of 1865.

The post-Civil War Reconstruction period is the most all-encompassing case of post-conflict stabilization in U.S. history. President Abraham Lincoln had wisely begun planning for postwar stabilization from 1863, developing mechanisms for reconciliation and infrastructure development projects to modernize the economy of the South, such as railroad and other infrastructure expansion to blunt Southern dependence on cotton once slavery was abolished. John Wilkes Booth's bullet brought much of that thoughtful preparation to a halt. Although not defined as "stabilization" at the time, historians suggest that the Army had learned applicable lessons[1] during the occupation of some Southern states while hostilities were still ongoing and even prior to the war in the recently acquired territories of California and New Mexico.[2]

WE DON'T DO NATION BUILDING!

Despite insistence from politicians that the United States does not "do nation building," the U.S. military has been engaged in this very activity nearly non-stop throughout its history, as illustrated by Fig. 1.1. Stabilization is a Joint[3] and Army[4] concept today and the most common activity for U.S. forces throughout their history.[5] It is also an essential step in translating military victory into long-term political success.

Yet despite these compelling facts, the U.S. government does not articulate the need for these activities, so the public does not support them.

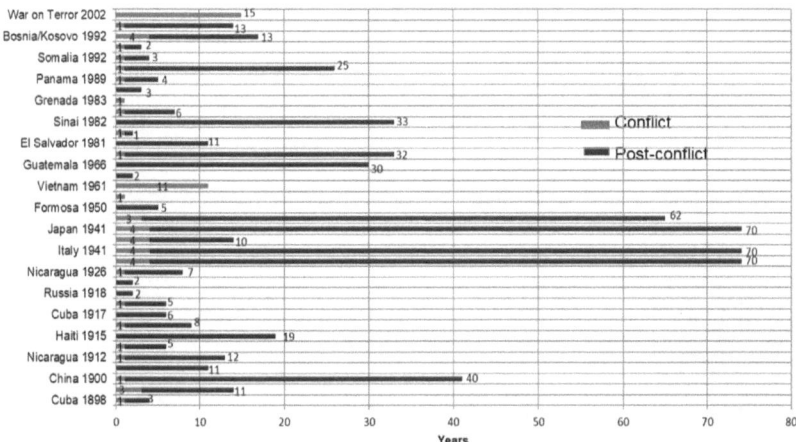

Fig. 1.1 Trend of long-term commitment in post-conflict conditions, 1891–2015

Thus, the military has to constantly rename what they are so the Congress and the people do not recognize that it has been deployed for anything other than combat. As a result, with notable exceptions—i.e., post-World War II Europe and Japan, and post-Korean War South Korea, stabilization is nearly always under-resourced and poorly planned, thus often failing to consolidate battlefield gains.

Although conducting war is what the Army trains and plans for, cleaning up after war, shoring up fragile states, and providing humanitarian assistance are the stabilization operations to which the Army as a tool is most often applied. Stabilization is defined in *Joint Doctrine (JP 3-07)* as "the process by which military and nonmilitary actors collectively apply various instruments of national power to address drivers of conflict, foster host-nation resiliencies, and create conditions that enable sustainable peace and security."[6]

Since 2009, the *Guiding Principles for Stabilization and Reconstruction*,[7] drawn from doctrine and practitioner expertise, serve as the stabilization "bible" for those in the field. Its *Strategic Framework for Reconstruction and Stabilization*, shown in Fig. 1.2, illustrates the six critical "End states" required for effective stabilization: Safe and Secure Environment, Rule of Law, Stable Governance, Social Well-Being, Sustainable Economy and the Cross-Cutting Principles without which the effort is expected to fail.

Fig. 1.2   USIP and PKSOI's stabilization lens

Most experts on stabilization agree that all of these elements are needed to varying degrees and in varying order to ensure stability in post-conflict or fragile states, but who should provide them, how, and to what degree for verifiable "end states" remains elusive. As political scientist Stephen Biddle noted in his 2013 article on the surge in Afghanistan, "Combat and security alone will have difficulty sustaining control if all they do is allow a predatory government to exploit the population for the benefit of unrepresentative elites."[8]

WHY RECONSTRUCTION?

This study is based on the post-Civil War Reconstruction period because it is the most significant case of post-conflict domestic stabilization in U.S. history. While stabilization was not yet a military area of expertise, some historians suggest that the Army had already learned valuable lessons conducting Reconstruction *during* the Civil War in parts of Arkansas, Louisiana, Tennessee, and Virginia. Army personnel had also gained experience drafting legal frameworks in the territories of California and New

Mexico recently acquired in the 1840s.[9] The U.S. Army gained additional experience in reconstruction in Mexico after the Mexican-American War, in which many Civil War generals participated as company grade officers.[10]

Reconstruction was the aftermath of a war fought at home and a peace still not enshrined in treaty a century and a half since cessation of large-scale violence. Louis DiMarco described this and its lessons for the future, thusly, "Post-conflict operations are a part of war; if the Army is to win the nation's wars, it must have a robust post-conflict capability. If the Army as an institution refuses to embrace this view of war and to plan and organize for it, it is doomed to repeat the failure of the occupation of the Confederacy."[11]

The August 2017 Department of Defense (DoD) *2016 Biennial Assessment of Stability Operations Capabilities (Biennial Assessment)* reflected DeMarco's assessment by stating, "Consolidating military gains into political victory requires stabilization efforts that enable legitimate actors to manage conflict peacefully – making stability operations an essential way for the United States to achieve its national security objectives."[12]

The *Assessment* acknowledges that "the role of DoD in stabilization efforts has changed over time throughout the course of U.S. history." It also notes that proof of effectiveness of the ongoing efforts in Afghanistan and Iraq is nearly nonexistent and the lack of planning for stabilizing the areas where the Islamic State of Iraq and Syria (ISIS) are no longer dominant, indicate the failure to understand the importance of stabilization in achieving sustainable U.S. national security objectives.

Defeated enemies do not just disappear, they regroup to fight another day, unless effective stabilization is conducted on a long-term basis. According to the second quarter 2019 DOD Inspector General report, the Islamic State has moved underground; retains 14,000–18,000 members; "is likely reestablishing financial networks" in both Iraq and Syria; "maintains an extensive worldwide social media effort to recruit fighters"; and has carried out asymmetric attacks using a "more stable" network for command-and-control and logistics.[13]

The U.S. military is suited to bringing security to any operational environment. The issues here focus on practitioners' and policymakers' nearly universal recognition that the Army is not the *appropriate* institution for most other stabilization tasks; but as it is the best *appropriated*, it is assigned operational tasking beyond its capabilities and capacity, leading to a pattern of stabilization failure. As noted in *Biennial Assessment*, the military prioritizes stabilization tasks and coordinates with other actors in accordance with *Framework* implementation by "**leading and conducting**" security, basic public order and providing for the immediate needs of the population, [while also] "**reinforcing and supporting** our interagency, international, and other partners in the priority areas of governance and economic stabilization to achieve national objectives."[14]

U.S. Reconstruction occurred domestically in an environment that was relatively permissive, with well-mapped terrain, basic infrastructure for maneuvering men and materiel, well established government institutions, and essentially shared language, culture and basic values, with some regional variation. This combination of beneficial factors is unique compared to foreign stabilization efforts, but the lack of understanding of the deepest culture of the South represented by the elite land-owning class that owed their power to slavery, caused Northern policymakers to underestimate the enormity of the task.

Analytic Method

Figure 1.2 presents the *Strategic Framework for Stabilization and Reconstruction*, which illustrates the *Guiding Principles' end states* or objectives and the *conditions* needed to achieve them. Within the conditions listed in white text in the figure there are additional *approaches* recommended for achieving them. All three levels within the five end states and the cross-cutting principles are presented as a series of matrices in *Annex I: Approaches, Conditions and End states for Stabilization and Reconstruction from the Guiding Principles*.

If the *Principles* are intended to serve as universal guidance, this *Framework* should be appropriate to any case of stabilization in any geographic location and at any time in history, although the actors may not yet have had the benefit of its codification. The stabilization case selected for this study is U.S. post-Civil War Reconstruction from 1863 to 1877.[15] This period was quintessentially about nation-building as it was a landmark time in which the United States attempted to (1) manage a post-conflict period, (2) shore up a fragile region, (3) integrate a long disenfranchised minority and (4) prevent future internal conflict.

The *Stabilization Assistance Review (SAR)*, signed by the Secretaries of Defense and State and the Administrator of the U.S. Agency for International Development (USAID) in 2018, considered the lessons from the *Biennial Assessment* and created a methodology to assess when the United States should appropriately engage in stabilization and a methodology for using all instruments and institutions of national power most effectively for each situation. The *SAR* defines stabilization as

> A political endeavor involving an integrated civilian-military process to create conditions where locally legitimate authorities and systems can peaceably manage conflict and prevent a resurgence of violence. Transitional in nature, stabilization may include efforts to establish civil security, provide access to dispute resolution, deliver targeted basic services, and establish a foundation for the return of displaced people and longer term development.[16]

The *SAR* Recommends the State Department and USAID take the lead in these efforts with Defense only taking on a supporting role. However, the budgetary resources of State are so dwarfed by the Defense budget that in reality the military ends up taking on the lion's share.

Of course, a fully comprehensive analysis of this case is far outside the scope of this brief study. Cringing historians aside, this analysis focused on key activities or events, whether they were undertaken by the military or other actors and how those activities contributed to achievement of the end states desired, mainly by Unionists, as expressed in this modern *Framework*.

To conduct this study, the researcher used a basic approach to evaluate the entirety of the *Framework* by converting each of the end states, the cross-cutting principles and their constituent conditions and approaches as illustrated in Fig. 1.2 into an MS Excel spreadsheet, available in *Annex I*. The next step was to note key actions that were taken during Reconstruction relevant to the recommended approaches.

The researcher assigned a basic score to each action: 1 to an action that had a positive outcome, −1 to a negative outcome and 0 for a neutral, mixed or uncertain outcome. It soon became apparent from the guidance for each approach that this was not an effective measuring method as the difference over time on whether an action had a "positive" or "negative" effect as intended was too significant to ignore.

As U.S. Army Lieutenant Colonel Joseph Long has noted in his study of insurgency leadership, when the risk is acute, certain decisive actions can be effective short-term that are unlikely to retain their efficacy in the long-term.[17] Therefore, the "outcome" column became two columns, one for "short-term" and one for "long-term" outcomes with the same scoring method, as illustrated in Fig. 1.3.

Basic Aggregate Assessment of Reconstruction Through Guiding Principles Lens				
	Reconstruction Effectiveness		Army Effectiveness	
	Short-term	Long-term	Short-term	Long-term
Cross-Cutting Principles	13	-14	8	-9
Rule of Law	23	-49	12	-23
Safe and Secure Environment	13	-26	6	-14
Sustainable Economy	26	-4	7	-3
Stable Governance	28	-23	20	-20
Social Well-Being	16	-13	9	-9
Overall	119	-129	62	-78

Fig. 1.3 Basic aggregated assessment of reconstruction through guiding principles lens

The short- and long-term scores are not equal for the two areas assessed as there were scores of "zero," indicating a neutral or unknown outcome. The results were unsurprising, but stark, as so much positive action toward reaching the intended end states took place in the short-term by all elements of society, strongly led by the Army, but much of it unraveled over the long-term. The "short-term" Reconstruction period lasted from 1863 to 1870, encompassing President Lincoln's initial rudimentary planning for postwar activities and ending when the majority of Union troops mustered out and withdrew from the former Confederate states. The "long-term" period lasted from 1870 to 1877, characterized by a Democratic-controlled Congress, the Compromise decision regarding the 1876 presidential election and the withdrawal of the remaining troops, which officially ended Reconstruction.

It must be noted that it is impossible to assess whether U.S. armed forces are the most effective lead instrument of national power for stabilization activities using the Army of 1867–1877. The nineteenth-century Army of Reconstruction was composed of short-term (3–6 months) civilians who had not ventured far from home except to go to this war and only had close-order drill and weapons training without any educational, doctrine or lessons learned system. They were not expecting occupation duty and desired to go home as soon as possible.

Today, the Army is composed of high school graduates at a minimum. It is a professional force expected to conduct post-conflict activities. It is supported by an extensive educational, lessons learned, doctrinal and provisioning system and has been deployed extensively around the world in differing situations.[18] Thus, a comparison of how the Army of Reconstruction would have handled the situation in today's Afghanistan or Iraq is not the goal of this study. Rather, the intent is to analyze the Cross-Cutting Principles and other stabilization and reconstruction "End states" to determine how the failure to address them at home has led to long-term lack of social reconstruction, thus exacerbating societies' challenges today and how to avoid the same result abroad in the future.[19]

NOTES

1. Mark L. Bradley, "The Army and Reconstruction, 1865–1877," part of the *U.S. Army Campaigns of the Civil War* series, Center of Military History, United States Army, 2015, available from http://www.history. army.mil/html/books/075/75-18/cmhPub_75-18.pdf. Accessed on July 20, 2017.
2. Mark L. Bradley, "The Army and Reconstruction, 1865–1877," part of the *U.S. Army Campaigns of the Civil War* series, Center of Military History, United States Army, 2015, p. 8, available from http://www.his tory.army.mil/html/books/075/75-18/cmhPub_75-18.pdf. Accessed on March 10, 2017.
3. Joint Publication 3-07 Stability, August 13, 2016, available from https://www.jcs.mil/Portals/36/Documents/Doctrine/pubs/jp3_07. pdf. Accessed on July 17, 2020.
4. Army Doctrine Publication (ADP) 3-07 Stability, July 2019, available from http://pksoi.armywarcollege.edu/default/assets/File/Doctrine/ ADP3-07_Stability_20190731.pdf. Accessed on July 17, 2020.
5. Lawrence A. Yates, "The US Military's Experience in Stability Operations, 1789–2005," available from https://web.archive.org/web/201 61029012226/http://usacac.army.mil/cac2/cgsc/carl/download/csi pubs/yates.pdf. Accessed on July 17, 2020.
6. Joint Publication 3-07 Stability, August 13, 2016, p. ix, available from https://www.jcs.mil/Portals/36/Documents/Doctrine/pubs/jp3_ 07.pdf. Accessed on July 17, 2020.
7. *Guiding Principles for Stabilization and Reconstruction*, United States Institute of Peace and United States Army Peacekeeping and Stability Operations Institute: Washington, DC, 2009, available from https://www.usip.org/sites/default/files/guiding_principles_ full.pdf. Accessed on August 6, 2019.
8. Stephen Biddle, "Ending the War in Afghanistan: How to Avoid Failure on the Installment Plan," *Foreign Affairs*, September/October 2013, available from https://www.foreignaffairs.com/articles/afghan istan/2013-08-12/ending-war-afghanistan. Accessed on July 7, 2020.
9. Mark L. Bradley, "The Army and Reconstruction, 1865–1877," part of the *U.S. Army Campaigns of the Civil War* series, Center of Military

History, United States Army, 2015, p. 8, available from http://www.his tory.army.mil/html/books/075/75-18/cmhPub_75-18.pdf. Accessed on March 10, 2017.

10. Noted by MAJ Kristan J. Wheaton (ret.) during manuscript review in March 2020.

11. Louis A. DiMarco, "Anatomy of a Failed Occupation: The U.S. Army in the Former Confederate States, 1865 to 1877," *Land Warfare Paper No. 66W*, November 2007, p. 1, available from https://www.ausa.org/publications/anatomy-failed-occupation-us-army-former-confederate-sta tes-1865-1877. Accessed on May 19, 2017.

12. U.S. Department of Defense *2016 Biennial Assessment of Stability Operations Capabilities*, Office of the Assistant Secretary of Defense for Stability Operations/Low Intensity Conflict (SO/LIC), August 2017, Washington, DC, p. 3.

13. Operation Inherent Resolve: Lead Inspector General Report to the United States Congress, April 1, 2019–June 30, 2019, p. 2, available from https://media.defense.gov/2019/Aug/06/2002167167/-1/-1/1/Q3FY2019_LEADIG_OIR_REPORT.PDF. Accessed on July 22, 2020.

14. U.S. Department of Defense *2016 Biennial Assessment of Stability Operations Capabilities*, Office of the Assistant Secretary of Defense for Stability Operations/Low Intensity Conflict (SO/LIC), August 2017, Washington, DC, p. 3.

15. "The Civil Rights Movement and the Second Reconstruction, 1945–1968," *History, Art & Archives: The United States House of Representatives*, available from http://history.house.gov/Exhibitions-and-Publications/BAIC/Historical-Essays/Keeping-the-Faith/Civil-Rig hts-Movement/. Accessed on August 25, 2017.

16. *Stabilization Assistance Review*, released jointly by Secretaries of the Defense and State Departments and the U.S. Agency for International Development Administrator on June 19, 2018, p. 4, available from https://www.state.gov/state-usaid-dod-stabilization-assistance-review-sar/. Accessed on August 5, 2019.

17. LTC Joseph Long, "Framing Indigenous Leadership," *Advances in Social Sciences Research Journal*, Vol. 4, No. 6, March 25, 2017.

18. Comments of COL Larry Bouchat (ret.) during review of an earlier version of this monograph received on June 3, 2020.

19. Caveats on comparing the two Armies were impressed upon the author by several colleagues from the U.S. Army War College in 2017.

CHAPTER 2

The Cross-Cutting Principles

Abstract *The Guiding Principles for Stabilization and Reconstruction* have served as a handbook for stabilizers since 2009 and provide a series of end state that must be addressed simultaneously to ensure sustainable peace after conflict. The "Cross-Cutting Principles" presented here are overarching and extend into each of the end states, sometimes considered the brain of stabilization, as they are indispensable and intractable from the other parts of the stabilization framework. These efforts are critical to influencing the leaders and the population itself to be empowered by stabilization and to take ownership of achieving the end states, so that peace is grafted indelibly into the indigenous social structure. These principles enable a whole-of-society effort toward transforming conflict into a positive force for growth, cohesion, and unity.

Keywords Political primacy · Legitimacy · Shared Understanding · Unity of Effort · Conflict transformation

Introduction

These principles are overarching and extend into each of the end states, sometimes considered the "grey matter" or brain of stabilization, as they are indispensable and intractable from the other parts of the *Framework*. These efforts are critical to influencing the leaders and the population itself to be empowered by stabilization and to take ownership of the various aspects of it so that it becomes sustainable and grafted indelibly into the indigenous social structure, thus enabling a whole-of-society effort toward a future with an ability to transform conflict into a positive force for growth, cohesion and unity.

CROSS-CUTTING PRINCIPLES: CONDITIONS AND APPROACHES

In the case of Reconstruction, failure resulted from a lack of planning for the postwar period and a lack of clear political vision among the two presiding executives and the legislature, which gave conflicting guidance to the military governors charged with enforcing the postwar peace. These failures to ensure **Unity of Effort, Political Primacy** and **Conflict Transformation** were also starkly apparent in Afghanistan, Iraq and Vietnam, as the strategy changed by political leader and from commander to commander, which precluded a consistent, unified approach to stabilization. If these underlying factors ensured failure in achieving the cross-cutting principles during a sustained domestic post-conflict effort, the process is highly unlikely to be effectively implemented abroad, even when the success of the post-conflict period is a priority to long-term U.S. interests (Fig. 2.1).

In particular as it relates to the Principles, a compelling question to explore is whether the South was a defeated internal insurgency requiring one form of "rehabilitation" prior to readmission to the Union or was it a defeated, i.e. conquered, sovereign rival needing "re-acculturation" prior to readmission to the Union? The northern Union never resolved that issue resulting in a lack of vision and mission clarity as to the purpose of Reconstruction and the role of the Army.[1] It also greatly contributed to the perpetuation of the "Lost Cause" myth still underpinning many Americans' views today, which suggests that the North won the war but the South won the peace.

One basic tenet of the Lost Cause is that the Southern secession was not prohibited by the Constitution, thus did not amount to treason. This would mean that the Confederate government and army were not traitors, but patriots fighting for states' rights as enshrined in the Constitution. True, this is part of the intent of a Federalist system, but their main reason

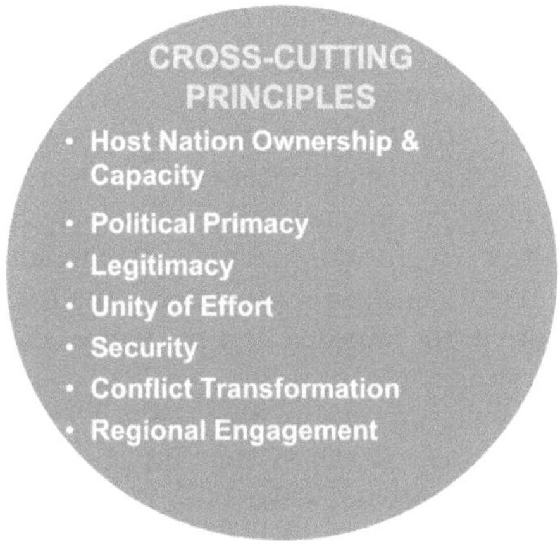

Fig. 2.1 Cross-cutting
principles

**CROSS-CUTTING
PRINCIPLES**

- **Host Nation Ownership &
  Capacity**
- **Political Primacy**
- **Legitimacy**
- **Unity of Effort**
- **Security**
- **Conflict Transformation**
- **Regional Engagement**

for insisting on the right of states to self-determination was to ensure the survival and expansion of states' to permit slave-owning. Another tenet of the Lost Cause myth is that these valiant patriots fought well and bravely under highly capable commanders and only lost the war because they were out-numbered and out-resourced by the greedy industrial North.

As further detailed in the section on *Stable Governance: Political Moderation and Accountability*, the failure of the North to recognize the power of women in preventing national reconciliation after reunion would be one of the biggest mistakes of the postwar period. To this day, in many circles, the myth of the Lost Cause affects our politics and shared ability to find national consensus in myriad areas and it is arguably a primary cause of the failure to integrate black Americans fully into the economic, governance, and social fabric of the country. While many Confederate and Union soldiers met for reunions and felt a desire for reconciliation through their share experiences on the battlefield during the Civil War and later in the Spanish-American War and the continuing battle to pacify the West.

Women from both sides did not have the occasion to mingle or share any common connections. Thus, Southern women, as thoroughly chronicled in Caroline Janney's 2013 *Remembering the Civil War*, kept the animosities leading up to the war alive and did the most to ensure the South remained "unreconstructed." In fact, with the official close of the occupation in 1877, Southern women of the next generation redoubled their efforts to focus the memory of the war on Reconstruction itself, not the Confederacy. They described the tyrannical federal government as occupiers who destroyed the "Southern way of life" and tried to "enslave white Southerners under governments led by incompetent, corrupt and racist black men, whenever the issue of black suffrage or equality of civil rights came up." This effectively canceled the intended effect of the postwar constitutional amendments ending slavery (13th), ensuring all citizens equal protection under the law (14th) and ensuring the right to vote is not impeded "on account of race, color, or previous condition of servitude" (15th).[2]

While the case of the U.S. civil war makes defining the host nation confusing, as the victor and absorptive government, the host nation for this purpose is the U.S. Federal government. By abruptly enforcing Federal demands of loyalty and slave emancipation, the Army failed to promote an historical form of **Host Nation Ownership and Capacity** in the Southern states as a way to build their capacity to govern for several years after the end of hostilities.[3] Federal troops were not completely withdrawn from the South until 1877; this de facto occupation resulted in hostility toward the Federal government that continues to linger and divide the nation.

This failure to recognize the adherence to the local slave culture of the regional elites, who would use all their remaining power to reverse any gains made during Reconstruction, especially in the area of civil liberties for freedpeople. They would also work to regain their lost land and create new power structures to ensure black labor continued to cultivate it.

ASSESSMENT

Although there was no handbook for military district commanders on how to administer their areas of responsibility precisely, most endeavored to work through local officials already in place, but found that they often had to replace mayors, magistrates, judges and police leaders for prejudicial behavior toward freed slaves and poor whites, as well as those who had opposed the war, known locally as *copperheads*.

In such cases, Army officers took on these roles to the best of their ability, but there was no clear effort to reeducate local officials on fairness and equality under the law or to build municipal capacity in this area, as the Army was already hard-pressed to do the work of maintaining an administrative and judicial system, let alone train existing officials on how to do their jobs according to Federal standards, which required the law to be applied equally after three new Constitutional Amendments were ratified (the 13th, 14th and 15th).

One critical approach to the condition of **Host Nation Ownership and Capacity** is the *role of women* and one that is often overlooked, although it is a key aspect of peacebuilding, especially after a conflict where many of the men in the region have been killed, exiled or otherwise removed from society. Women are left to manage many of the responsibilities once held by men, which can often result in profound societal changes, as have been chronicled in post-World War Europe and in places suffering long civil conflict, such as Liberia.

Although the Army of the 1860s did not have the same understanding of gender issues that we do today, it recognized the special contributions made by women on both sides during the war and its aftermath. In 1861, the Federal government established a Sanitary Commission that by 1865 had distributed $15 million in supplies to ensure the hygiene of medical and food provisions, nearly all of which had been collected by women. According to Civil War historian, Dr. Raymond Millen, these efforts convinced many soldiers of the organizational capabilities of women, forcing many to recognize they were at least equal to men in such endeavors, thus dispelling the myth of their inferiority.[4]

The 3000 women who had served as nurses or hospital administrators or many more who had to run their own farms and households during the war often felt empowered and when the men returned, they did not always want to abdicate their new roles. In addition, hospitals still had many casualties to care for and needed nurses and administrators. Many men did not want their wives working outside the home once they returned from war, which created additional rifts in society.[5]

Just as after World War II, many women had worked in munitions and textile factories to help the Northern war effort. Many other women, such as Cady Stanton, Susan B. Anthony and Harriet Beecher Stowe, whose *Uncle Tom's Cabin* is seen by some historians as a slow trigger for the war,[6] were deeply involved in politics. We also know of over 400 women who disguised themselves as men to fight with either the Confederate or Union armies and many more served as spies in various capacities.[7] It would prove difficult to send these back to the kitchen once the fighting stopped.

After the war, many female nurses and teachers from New England went south to help provide healthcare and education for the estimated four million newly freed people under the auspices of various charitable groups as well as the Congressionally created U.S. Bureau of Refugees, Freedmen and Abandoned Lands, commonly called the Freedmen's Bureau.

Finally, it is difficult to discuss postwar reconstruction and stabilization in any era or location without considering the role of women in peace and security. The U.S. Congress passed the "Women, Peace and Security Act," in September 2017 expressing that the United States recognizes the critical role of women "in fragile environments in conflict prevention and conflict resolution [to] promote more inclusive and democratic societies [for] country and regional stability."[8]

This role was often discounted after the Civil War, although General Carl Schurz declared, in 1865, after seeing a Southern woman in a hotel restaurant furiously refuse to take a dish of pickles from a Union officer's hand, that women were "a hostile moral force of incalculable potency."[9] Confederate women supported the rebellion with food, clothing and safe haven, by writing letters and petitions and sent their sons to war proudly and sometimes forcibly. Women were also at the forefront of keeping national divisions alive with a flurry of Confederate monument building in the immediate postwar period. Historian James McPherson even quipped, "If the Confederacy had raised proportionately as many soldiers as the postwar South raised monuments, the Confederates might have won the war."[10]

As Caroline Janney's 2013 book, *Remembering the Civil War* painstakingly recounts, southern women were at the forefront of establishing and maintaining Southern military cemeteries and developing the protocols for memorial day observances later co-opted by the Northern states as "Federal" holidays. Southern women organized political rallies and ensured the Southern view of the war would be remembered by future generations, going so far as to establish textbook commissions that sought to remove all references to Southern states and statesmen as "rebels," "traitors" or "insurgents." Women such as Mildred Rutherford with her book, *A Measuring Rod to Test Text Books…in Schools* ensured that the country reunified but to this day has not necessarily reconciled. The Southern myths of the Lost Cause and insistence that Americans of African descent are less intelligent or capable than Caucasians, ensured black Americans would be disenfranchised and disempowered.

The *Guiding Principles* espouse the basic premise of **Political Primacy** in post-conflict environments, meaning "every single aspect of stabilization and reconstruction will have a political element to it."[11] Politics, no matter how it the process or institutions are organized, is essentially a struggle for power. Even though the hallowed democratic process is meant to empower "the people," some people will end up with more power than others. The aim of the so-called "Radical" Northern Republicans at that time was to utterly upset the power balance tipped toward the Southern elites by empowering black and non-landowning white people.

This was no less true in the 1860s and 1870s than it is today in the Middle East and Africa. In those latter places, much effort is devoted to developing democratic institutions that mirror Western democracy, as this "government by the people" is considered by donors to be the most humane and inclusive thus developed. However, often the sources of power are resources, as in the case of the American South, this translated to land and labor. Some 2000 plantation owners who comprised the Southern oligarchy controlled the majority of the land, the labor, and thus the politics and the path to wealth accumulation with little to no representation from poor white farmers.

Often, elites maintain their grip on power while adjusting electoral and other laws that benefit the status quo. This is precisely what happened in the American South after the Army withdrew in 1877. The "Redemption" period followed Reconstruction as the same Democratic politicians regained their hold on Congressional seats and Governorships as well as local official positions.

Most historians who acknowledge Reconstruction as a failure, point to the staggering efforts the Army and its commanders expended to safeguard emancipation and ensure security but agree to varying degrees that it was the political objectives of the winning side, namely Northern Republicans, which were not sustainably achieved. Primary among these objectives was the abolition of slavery and granting of full citizenship rights to former slaves, including the right to vote and hold elected office. Some historians call these goals self-serving, as the vast majority of freedmen were likely to exercise their new rights and vote and run as Republicans, thus instantly adding over a million members to the party's rolls.

In 1938, historian Robert Selph Henry characterized Reconstruction after President Abraham Lincoln's assassination as a "long-drawn-out tragedy of misunderstanding."[12] While this Southerner's view may have been clouded by his recent memory of an even larger tragedy caused by misunderstanding—World War I—the description is apt. Lincoln's genius was in his flexibility, even refusing to sign a bill codifying his own plan for national reconciliation on the grounds that it was only one plan and perhaps another, even better one, could be conceived but would be impossible to implement if the foundation of the effort were cemented by the first plan. Lincoln's political acuity extended to great empathy for the Confederate position, recognizing it as an economic, rule of law, and cultural issue.

When looking at Reconstruction as a stabilization effort seen through the lens of the modern *Strategic Framework*, it becomes apparent that the issue of slavery was the primary impediment to conflict transformation, political primacy and legitimacy, as it permeated all aspects of Southern society. Slavery is related to all of the *Framework's* end states, for the majority of Southerners of the time felt slavery was essential to their economy. The institution also represented their right to self-government (its abolition validating their pre-war reasoning for secession) and was endemic to their view of an orderly society.

As Dr. Millen has pointed out, "Slavery had been a divisive issue for years with politicians striving to reach peaceful solutions – the Missouri Compromise, the Compromise of 1850, and others. Although few historians acknowledge it, the annexation of new territories/states made the issue of slavery irreconcilable. The war would have occurred sooner or later because of the electoral implications."[13]

The **Legitimacy of the Federal government** as a critical factor of stable governance was in question long after the end of the war. While the issue of slavery has become quite blurry over the past 150 years post-Civil War with many revisionist scholars claiming that the war was about other things, economic or political issues, this apologist takes entirely lacks utility in any analysis of the postwar period as slavery was the primary cause of the war. As effectively portrayed by the memorable characters in the 1939 film *Gone with the Wind*, the Army's enforcement of abolition and the initial enforcement of former slave enfranchisement dealt a severe blow to many Southerners' ability to find legitimacy in the united government, seeing Reconstruction as an occupation and a force bent on destroying what was left of Southern life.

Not only did most white Southerners see black people as inherently destined to be slaves, but a common interpretation of a passage in the *New Testament Book of Genesis* Chapters 9–11 suggested Noah's son Ham was cursed to be the "servant of servants." Later editions describe Ham as "black," and finally even of African descent. Historians have identified literature from as early as the 1670s using the "curse of Ham" as white justification for—and abolitionist attack against—the institution of African enslavement.[14]

Later American religious works describe Ham as a lewd and disreputable son to Noah and treat his descendent, King Nimrod of Babel, as a sort of insurgent leader who brutalizes his people, takes others' property and territory and dishonors God. Even the derivative word "nimrod" has been transmogrified from its original sixteenth-century meaning of "a skillful hunter" to the later North American usage meaning "an inept person." This entry also notes that the word derives from Noah's great grandson, a noted hunter.[15] Interestingly, the British Navy has had six ships called the *Nimrod*, including the one Ernest Shackleton sailed on his 1907–1909 Arctic expedition. Apparently the word is not viewed in English the same way it was construed in "American."

Such characterizations led easily to proliferation of stories of inherent ignorance and laziness and the mocking and underhanded Uncle Remus stereotypes in American literature as well as the "White Man's Burden" philosophy used to justify Western imperialism, which all made black slavery seem a kindness to people unable to govern or discipline themselves.[16] These cultural memes enabled both the Southern white population and the enslaved black populations to see black people as inferior, thus justifying continued enslavement.

Anti-abolitionists even tried to make the case that emancipation was unkind, making former slaves work for poor wages and then pay room and board, so it was more compassionate to leave slaves in the "care" of their masters where they belonged and would be housed, fed and clothed in Christian charity. In 1861, the sermon of one minister in Georgia was reprinted in the *Macon Telegraph*, declaring

> Both Christianity and Slavery are from heaven; both are blessings to humanity; both are to be perpetuated to the end of time…Slavery is right, and because the condition of the slaves affords them all those privileges that would prove substantial blessings to them, and too, because their Maker has decreed their bondage, and has given them, as a race, capacities and aspirations suited alone to this condition in life.[17]

Radical Republicans in the North intended that once freed, former (male) slaves should have all the rights of citizenship enjoyed by white men immediately following the war, while the Southern Democratic position was polar opposite. This sharp divide in their views of slavery was a touchpoint that lingered as black enfranchisement failed to become a reality until 100 years after the end of the Civil War and only then after renewed reigns of terror in the South. The pervasive view that black Americans are "unable to govern themselves, thus they should not be permitted to govern others through their votes."[18] This was another big part of the Lost Cause myth of "benevolent masters and faithful slaves" that have haunted African Americans to this day, not only in the South, but the majority of white Unionists believed that they were "socially, culturally, physically, and mentally superior to African Americans."[19]

During the occupation, Army commanders typically tried to maintain the **Unity of Effort** and although the concept of the interagency or a "whole of government" did not exist in 1865, the Army quickly understood that it did not possess all the capabilities it needed to effectively carry out its mandate and would have to contract for these skills. Contractors included lawyers serving as magistrates and judges, as well as advocates for the government itself and for freedpeople and other civilians in disputes, and teachers, doctors and nurses to provide education and healthcare to freedpeople as they flocked to the food, shelter, safety and mounting services offered by Army posts cum internally displaced persons (IDP) camps. Today, the U.S. Agency for International Development (USAID) among other nongovernmental and international organizations, would provide such services to a Host Nation but the Army essentially had to go it alone during Reconstruction, predating today's Federal Emergency Management Agency (FEMA) for domestic disasters.

From the very start of the occupation, there was a lack of **Shared Understanding of the Situation**, one of the crucial approaches to establishing Unity of Effort in the field of battle or stabilization. Both the Federal Executive and Legislative branches seemed to underestimate the magnitude of the ideological fervor and deep resentment against the vanquishing/occupying force. General Ulysses S. Grant traveling in the North immediately after the war, for instance, had a greatly divergent view of the level of threat from that of his commanders in the Southern Districts.

The issuance of Second and Third legislature-sponsored Acts superseding (intentionally) Johnson's Presidential Reconstruction instructions, left district commanders flummoxed as to how to serve the letter and spirit of the legislature at the displeasure of their Commander in Chief. This confusion between Congressional legislation conflicting with Presidential policy led to an untenable position for many commanders, some of whom Johnson replaced by 1867.[20]

In addition to the lack of a formal peace agreement to guide postwar activity and ensure **Security**, the disarmament, demobilization and reintegration (DDR) process was not organized nor consistent. As historian Gregory Downs sagely stated, "Into the vast, unknown space of the South would soon march hundreds of thousands of rebel soldiers heading home. Young, isolated, unpaid, angry, and well-armed, these white men could easily join uprisings or launch guerilla wars."[21]

Removing the word "white" makes this sentiment applicable to the youth in nearly any postwar society in any era. After the fall of Muammar Ghaddafi, thousands of ethnic Tuareg mercenaries returned home from Libya to Mali and Niger, armed and battle-seasoned from service in Libya to find poor governance, lack of economic opportunity and marginalization, especially in northern Mali, sparking a conflict that resulted in a coup and a United Nations stabilization mission deployment to the region in 2013. To date, this mission has been the deadliest in UN history with 142 killed as of June 30, 2020.[22]

In this light, Downs' further comment is even more familiar, "civil wars often sparked spiraling conflicts, echo wars that sounded for years or for generations as ex-soldiers continued to strike back against the victors."[23] Enter the ISIS in Iraq or al Qaeda in the Islamic Mahgreb (AQIM) in Mali today, who initially swept up the returning secular Tuaregs who sought independence on nationalist and economic grounds into its own ideological struggle to impose its stated "way of life" under Sharia law.[24]

Many practitioners advocate extending DDR programs beyond ex-combatants to any significantly marginalized group in a society not healed after war. Failure to fully implement the 14th and 15th Amendments granting rights to freed black men in the United States, the practice of sharecropping and enforcement of local vagrancy laws enacted immediately after the cessation of violence led to states' expanding legal restrictions on the movements and opportunities of black citizens once many rejoined the Union, which essentially maintained the vestiges of slavery if not the name. Culminating in implementation of a collection of laws in Southern states limiting the rights of black Americans by claiming to keep the races segregated while providing anything but "equal" access to the law, education, transportation and virtually all public services, collectively referred to as "Jim Crow" laws enacted in the 1890s, remained in place until after the enactment of the 1964 (second) Civil Rights Law.

Jim Crow referred to a derogatory stereotype of a fawning, clumsy, ignorant black caricature culled from minstrel shows that developed into a series of segregating and disempowering laws enacted across the South to ensure black Americans were unable to exercise the rights granted under the 13th, 14th, and 15th Constitution amendments. They disallowed blacks from accessing schools, hospitals, public transportation, and other public amenities and greatly curtailed economic opportunities and even basic civil rights until they were rolled back in the middle of the 1960s. This system essentially evolved into a caste system that even included lynching and other forms of violence systematically perpetrated mainly against black men.[25]

According to the *Guiding Principles*, **Conflict Transformation** is about reducing the drivers of conflict and supporting the growth of conditions that undergird stability and development in the long-term.[26] In the Civil War, the primary driver of conflict was the political power struggle to protect elites' hold on land and labor, centering on the issue of slavery, which was abolished in theory with the Emancipation Proclamation and finally in fact through Constitutional Amendments and by force through the Army's occupation and associated activities for the next dozen years postwar. Throughout this process, some historians argue that the war continued, even as "battlefield deaths" were curbed, a rise in "political deaths" continued. Civil War historian Dr. Mark Grimsley has noted that war is defined as "an event that results in 1000 or more deaths each year." Under this definition, he makes the case that Reconstruction constituted a "second civil war."[27]

While the widespread violence brought on by the war ended at Appomattox in April 1865, the South and the Army had very different ideas of what postwar "peace" should look like. Since there was never a peace agreement officially ending the war, the wartime conditions were permissible under law and it can easily be argued that the war still has never ended. As the individual Southern states worked through the political process of regaining home rule at varying velocities, in some places politics took on the characteristics of a traditional insurgency. Stephen Budiansky recounts that from 1865 through 1877, "more than 3,000 freedmen and their white Republican allies were murdered in the campaign of terrorist violence…[including] more than 60 state senators, judges, legislators, sheriffs, constables, mayors, country commissioners, and other officeholders whose only crime was to have been elected."[28]

Notes

1. As stated by Dr. Andrew Roth during a review of the manuscript in March 2020.
2. Fifteenth Amendment to the U.S. Constitution, available from https://www.ourdocuments.gov/doc.php?flash=true&doc=44. Accessed on July 20, 2020.
3. Notwithstanding the fact that the South did not invite the Army in, as would be required in an international intervention, the HN concept is understood at its basic level here.
4. Dr. Raymond Millen, Civil War historian and Professor of Security Sector Reform at the U.S. Army War College added during peer review December 2017.
5. "Women in the Civil War," The History Channel, A&E Television Networks, LLC, 2017, available from http://www.history.com/topics/american-civil-war/women-in-the-civil-war#. Accessed on August 25, 2017.
6. Dr. Raymond Millen, Civil War historian and Professor of Security Sector Reform at the U.S. Army War College during peer review December 2017.
7. "Women in the Civil War," The History Channel, A&E Television Networks, LLC, 2017, available from http://www.history.com/topics/american-civil-war/women-in-the-civil-war#. Accessed on August 25, 2017.
8. From text of S.1141—Women, Peace, and Security Act of 2017, passed by the 115th Congress on August 3, 2017, available from https://www.congress.gov/bill/115th-congress/senate-bill/1141. Accessed on October 10, 2017.
9. Richard White, *The Republic for Which It Stands: The United States During Reconstruction and the Gilded Age, 1865–1896*, Oxford History of the United States Series, Oxford University Press: Oxford, UK, 2017, p. 26.
10. Historian James McPherson quoted by Lawrence A. Kreiser, Jr. and Randal Allred in *The Civil War in Popular Culture: Memory and Meaning*, University Press of Kentucky: Lexington, KY, 2014, p. 67.
11. *Guiding Principles for Stabilization and Reconstruction*, United States Institute of Peace and United States Army Peacekeeping and Stability

Operations Institute: Washington, DC, 2009, pp. 3–15 and 3–16, available from https://www.usip.org/sites/default/files/guiding_principles_full.pdf. Accessed on August 25, 2017.

12. Robert Selph Henry, *The Story of Reconstruction*, 1999 reprint. Konecky & Konecky: New York, NY, p. 5. Originally published in 1938 by the Bobbs-Merrill Company.

13. Dr. Raymond Millen in conversation with the author on November 13, 2017 at the U.S. Army Peace Keeping and Stability Operations Institute, U.S. Army War College, Carlisle Barracks, Carlisle, PA.

14. Stephen R. Haynes, *Noah's Curse: The Biblical Justification of American Slavery*, Oxford University Press: London, UK 2002, pp. 8–10.

15. Dictionary.com entry for "nimrod," Available from http://www.dictionary.com/browse/nimrod?s=t. Accessed on August 28, 2017.

16. Stephen R. Haynes, *Noah's Curse: The Biblical Justification of American Slavery*, Oxford University Press: London, UK 2002, pp. 8–10.

17. From "Scriptural Vindication of Slavery," Sermon by the Reverend Ebenezer W. Warren, Minister of the First Baptist Church of Macon, Georgia delivered on January 27, 1861; cited in Bruce T. Gourley, *Diverging Loyalties: Baptists in Middle Georgia During the Civil War*, Mercer University Press: Macon, Georgia, 2011, pp. 19–20, available from Google Books. Accessed on July 6, 2017.

18. Quoted from *New York Times*, May 13, 1899 articles in Caroline Janney, *Remembering the Civil War: Reunion and the Limits of Reconciliation*, University of North Carolina Press, 2013, p. 228.

19. Caroline Janney, *Remembering the Civil War: Reunion and the Limits of Reconciliation*, University of North Carolina Press, 2013, p. 200.

20. Mark L. Bradley, "The Army and Reconstruction, 1865–1877," part of the *U.S. Army Campaigns of the Civil War* series, Center of Military History, United States Army, 2015, p. 31, available from http://www.history.army.mil/html/books/075/75-18/cmhPub_75-18.pdf. Accessed on March 10, 2017.

21. Gregory P. Downs, *After Appomattox: Military Occupation and the Ends of War*, 2015, Harvard University Press: Cambridge, MA, p. 13.

22. This figure excludes those killed by Accident (28) or Illness (49). United Nations Multidimensional Integrated Stabilization Mission in Mali (MINUSMA) "History" webpage, available from https://minusma.unmissions.org/en/history. Accessed on August 5, 2019. Death toll data, available from https://peacekeeping.un.org/sites/default/files/stats_by_mission_incident_type_4_52.pdf. Accessed on July 22, 2020.

23. Gregory P. Downs, *After Appomattox: Military Occupation and the Ends of War*, 2015, Harvard University Press: Cambridge, MA, p. 13.

24. Diane E. Chido, *Intelligence Sharing, Transnational Organized Crime and Multinational Peacekeeping*, June 2019, McMillen & Co. Palgrave

Pivot, p. 23, available from https://www.palgrave.com/us/book/978 3319711829. Accessed on January 28, 20120.

25. "What Was Jim Crow," Jim Crow Museum of Racist Memorabilia, Ferris State University, 2012, available from https://www.ferris.edu/jimcrow/ what.htm. Accessed on July 22, 2020.

26. *Guiding Principles for Stabilization and Reconstruction*, United States Institute of Peace and United States Army Peacekeeping and Stability Operations Institute: Washington, DC, 2009, pp. 3–21, available from https://www.usip.org/sites/default/files/guiding_principles_ full.pdf. Accessed on August 25, 2017.

27. Mark Grimsley, "Wars for the American South: The First and Second Reconstructions Considered as Insurgencies," *Civil War History*, Vol. 58, No. 1, March 2012, pp. 6–36.

28. Stephen Budiansky, *The Bloody Shirt: Terror After the Civil War*, 2009, Plume Penguin Publishing: New York, NY, pp. 7–8.

CHAPTER 3

Stabilization and Reconstruction End States

Abstract As the Army tends to plan strategically for long-term success in complex engagements, troops on the ground are tasked with series of shorter-term, outcome-focused missions. The objective is to complete the mission successfully and return for the next tactical order to complete operational requirements that will lead to overall strategic victory. As effective as this approach may be for effective combat success, it is critical to recognize the incompatibility of this approach for stabilization.

Keywords Security · Governance · Rule of law · Economic sustainability · Social well-being · Freedmen's Bureau · Women in peace and security

Introduction

As the Army tends to plan strategically for long-term success in complex engagements, troops on the ground are tasked with series of shorter-term, outcome-focused missions. The objective is to complete the mission successfully and return for the next tactical order to complete operational requirements that will lead to overall strategic victory. As effective as this

© The Author(s), under exclusive license to
Springer Nature Switzerland AG 2021
D. E. Chido, *US Army's Effectiveness in Reconstruction According to the Guiding Principles of Stabilization*,
https://doi.org/10.1007/978-3-030-60005-1_3

approach may be for effective combat success, it is critical to recognize the incompatibility of this approach for stabilization.

Military operations other than war (MOOTWA), as stabilization is often called, require a strategically minded, long-term approach that the Army is not structured to or accultured to conduct. As noted earlier, the Army tends to be selected for stabilization because it is the best resourced and because the military will agree to undertake whatever missions it is assigned, no matter how messy or complex. As also noted before, stabilization requires not only a longer-term mindset, but a more diplomatic approach using more of the instruments of national power beyond force to be effective. But, as the Federal budget is unlikely to be rebalanced away from the massive funding for Defense, the military is likely to continue to be the primary tool for this job, as opposed to only supporting it as recommended in the SAR and elsewhere.

Thus, the idea of stabilization end states, are ephemeral, as these situations are dynamic and can literally shift with the weather, as a result of a natural disaster, or with a change in U.S. or Host Nation political leadership. Thus, the longer-term approach and commitment to ensuring a sustainable Safe and Secure Environment, Rule of Law, Stable Governance, Economic Development, and Social Well Being will continue to be called end states in the *Strategic Framework*. This chapter will discuss each of them as represented by their colored circle from the graphic presented in the introduction, that also includes the conditions needed to reach those end states. As indicated in the spreadsheets presented in *Annex I: Approaches, Conditions and End States for Stabilization and Reconstruction from the Guiding Principles*, readers can see the full set of approaches taken to achieve each condition in order to support end state sustainment. The chapter will explore these as they relate to post-U.S. Civil War Reconstruction.

Safe and Secure Environment

While the **Cessation of Large-Scale Violence** was greeted with relief on both sides, the ineffectual approach to "winning the peace" quickly became apparent, with the failure to ratify or even draft a peace agreement. This prevented the Reconstruction process from having an initial blueprint to which both sides had agreed, creating no will for the "defeated" South to make the changes required to reenter the Union, have common expectations with which to align themselves with a shared future and make the societal changes that were mandated by military-led martial law that retained war powers indefinitely (Fig. 3.1).

Fig. 3.1 Safe and secure environment

As noted in the *Strategic Framework*, the second necessary condition for ensuring a safe and secure environment is an **Enduring Cease-fire or Peace Agreement**. The failure of both sides to share a peace agreement is considered by some historians an important hindrance to healing Southern pride after the military defeat and the losses that the end of slavery and the prospect of the elevation of former slaves to full citizenship would engender.[1]

The lack of a peace agreement made the second of the critical approaches toward the conditions needed to achieve a **Safe and Secure Environment** also impossible to achieve, at least in the short-term. The recommendation to transform the conflict to pursuit of political and economic goals by nonviolent means was undermined by the Union's unconditional requirement that Southern states adhere to its formula for readmission to the Union with no consideration of how this might affect Southern states' citizens.

Assessment

The *Biennial Assessment* defined "security," the primary role of the military in stabilization operations, as "A terrain-based activity to provide internal security such that borders are protected, and people and goods can move freely throughout the country…the bedrock condition that creates a sufficiently peaceful environment that permits the other stability functions and tasks to be carried out. Stabilization must also extend security requirements to the protection of the civilian population from violence."[2]

As the number of troops available for occupation of the South constantly dwindled, there was never an effective or widespread disarmament process; Southern officers retained their weapons and soldiers surrendered their rifles in some areas and were permitted to retain their personal weapons to protect their families and property in others, but even this level of disarmament did not take place universally, ensuring there were plenty of arms available to enact campaigns of violence and intimidation against freedmen and sympathetic whites once the Army receded.

Another contentious issue that prohibited securing a final peace was ambiguity in the Constitution over whether the President or the Congress was responsible for establishing the peace. Congress possesses the power to declare war and Senate ratification of treaties is required for them to become acted upon. In the case of civil war, it is unclear who had the right to declare peace. This was not resolved as both Presidents Lincoln and Johnson insisted that they had the power to maintain a state of war in order to use the Army to enforce Reconstruction, while Congress insisted that once all Southern states were again represented in the House and Senate, peace would be achieved. According to Army code, martial law remained in place until the Commander in Chief declared it over.[3]

In order to reestablish full representation, President Lincoln issued the so-called "Ten Percent Plan," which dictated that following cessation of hostilities, once 10 percent of the 1860 electorate in a given Southern state took the oath of loyalty to the Union and denounced slavery, that state could elect delegates to a convention to rewrite the state constitution reflecting loyalty and emancipation. Once this was passed by the convention, the state could then elect a legislature and the military government would withdraw. By this measure, peace was not achieved until a Georgia senator was seated in Washington in February 1871.[4]

In the absence of a peace agreement, generals responsible for particular geographic districts agreed sometimes verbally with their opposite numbers to certain courses of action (COAs) based on their personal codes of honor to establish **Legitimate State Monopoly over Means of Violence**, and **Physical and Territorial Security**. For instance, in their area of operation (AO), once Union General William Tecumseh Sherman verbally assured General Joseph Johnston in April 1865 that the South would not become subservient to the North and denied representation in Congress, Johnston agreed to end hostilities. Under this agreement, Confederate soldiers were ordered to return to their states, remanding their arms to the state arsenal in their respective capital. Once this had been done and officers had sworn an oath of allegiance to the Federal government, it would recognize the legitimacy of the state government. This agreement did not conform to the Ten Percent Plan yet to be promulgated.[5]

The Second Reconstruction Act divided the southern states into five military districts to register voters, monitor elections, and convene a constitutional convention in each state to pass a new constitution guaranteeing suffrage for freedmen. To manage eight million people over an area the size of Western Europe, a total of 20,000 soldiers were deployed with the end date only designated in each state once a constitution was ratified, with no clearer timeline specified.[6] Once again, lack of adequate resources reduced the directed authorities' ability to conduct effective stabilization activities. After one particularly violent skirmish in Mobile, Alabama, Secretary of War Edwin Stanton instructed district commanders to focus troop strength on cities and towns to maintain visibility and increase mobility.[7]

As other states were gradually readmitted to the Union, Georgia remained an outlier. Various Ku Klux Klan groups were extremely active in the state, inspiring terror and engaging in criminal activities, with violence against blacks and sympathetic whites. By August 1869, Commander of the South, Brigadier General Alfred H. Terry, recommended rescinding Congress' June 1868 invitation to rejoin the Union and that Congress impose stricter requirements for readmission. In January 1870, President Grant declared the state a military district and put Terry in command as had been done during the initial postwar period.

Although some of Terry's actions were deemed illegal at the time, such as removing 19 legislators on tenuous grounds of "ineligibility," Klan activities were severely curtailed, and Georgia became more peaceful. By July 1870, Georgia was readmitted, and the Union regained all of its territory in law if not in sentiment.[8]

At the same time, Klan elements had effectively taken control of two counties in South Carolina. After several violent acts committed against public figures, Governor William Holden declared these counties to be in the throes of insurrection and called up the state militia, while appealing to President Grant for Union troops to help establish order and lawful-ness.[9] At this time, Union troops, particularly cavalry, were spread thinly in the South as they were fully engaged in Western expansion, thus remaining Southern units were relegated to chasing the Klan on foot, which gave the "Night Riders" an immense advantage.

Recognizing this disparity and the lack of necessary strength due to the March 1868 Manpower Reduction Act, Grant was more than amenable to assist and in addition to pledging to send Federal troops, he even offered to equip the state militias. As a result of this troop "surge," Klan activities ceased almost immediately, which had two consequences: the disbanding of the militia and return of all North Carolina territory to home rule, but the threat of Klan terror still kept Republican party voters away from the polls, which led to a solid Democrat majority on Election Day August 4, 1870.[10]

Although the postwar nation has remained unified and essentially stable for over a century and a half, this sectarianism has persisted in the United States and manifests today in the form of economic and security dispari-ties for black Americans, although they were ostensibly granted full rights of citizenship after the Civil War and again over 50 years ago. Racially motivated violence has also continued sporadically but recent populist rhetoric and the ubiquity of real-time "news" have suggested it is on the rise. A review of the post-Civil War reconciliation process brings to mind the kinds of sectarian civil violence observed today in Afghanistan, Iraq, Libya, Mali, Nigeria, and Syria. These are all places where no final peace has been agreed and varying views of how society should be organized to best serve each faction results in endless violence.

One of the approaches to the **Safe and Secure Environment** end state in the *Guiding Principles* is to "promote the civil authority of the state, as long-term stability depends on it."[11] Unfortunately, as noted throughout this assessment, the political, judicial, and security apparatus was strongly prejudiced against free blacks and poor whites, as well as those who had opposed the war, so Army commanders often replaced local officials and provided security, law enforcement, and military courts.

The Army endeavored to protect the rights of blacks and poor whites by educating them on their rights, running military courts, and providing legal services using their own officers and contractors and by assisting agents of the Freedmen's Bureau. As the Army's resources in the South were reduced and states rejoined the Union, responsibility for safety and security was gradually returned to local officials, often the same ones replaced in early Reconstruction with no true changes in the legal codes to protect freedpeople.

Rule of Law

The Federal government endeavored to provide **accountability** when local justice proved prejudicial toward freed blacks and poor whites. Freedmen and loyal southern whites descended upon Army posts and Freedmen's Bureau agents for legal protections and dispute resolution. As the military withdrew, legal service access was curtailed to areas closest to posts and in the largest cities, resulting in unfortunate outcomes such as the growth of the convict lease program and the loss of civil rights such as the right to property and participation in legal proceedings and politics as "Jim Crow" laws expanded (Fig. 3.2).

Instead of international engagement from modern institutions such as the United Nations (UN) leading peace keeping and stabilization operations, the U.S. Federal government was considered "foreign" in much of the South and the Union officers often considered themselves occupiers of a hostile "country."

Fig. 3.2 Rule of law

RULE OF LAW

• Just Legal Frameworks

• Public Order

• Accessibility to the Law

• Access to Justice

• Culture of Lawfulness

ASSESSMENT

The *Biennial Assessment* refers to the 2016 *JP-3.07*, which states that "Transitional public security is a specific requirement that extends the role of the military beyond its traditional roles and missions." Further, the *Assessment* asserts

> As an interim measure, U.S. forces may have to conduct transitional public security (TPS) and assume responsibility for maintaining public order in the place of host nation police forces. The purpose of TPS is to protect civilian populations when the rule of law is broken down or non-existent. TPS includes a spectrum of activity, from protecting civilians and property from violence to building large-scale law enforcement, judicial and correctional systems. As the responsibilities for public security transition from DoD to competent civilian authority, military efforts support police and judicial reform led by the competent civilian authority.[12]

According to the *Guiding Principles*, **Just Legal Frameworks** include "a constitution, legal codes, acts, decrees, binding regulations, bylaws, standard operating procedures, case law, peace agreements, Status of Forces Agreements [SOFAs]...that are...legally certain, drafted in a transparent way, publicly promulgated, and ensure the separation of powers, including judicial independence." They emphasize that "Just laws are also fair, equitable, responsive to the needs and realities of the host nation and benefit the entire population, not just powerful elites."[13]

This underscores the problems of stabilization as even development of new, transparent legal frameworks and institutions do not guarantee stability without security to ensure they are firmly established and can function according to the international standards under which they are created. They must also have indigenous cultural bases in order to be perceived as legitimate to those they intend to govern, which can be at odds with international norms. At the same time, security forces must remain in sufficient volume and capability to ensure institutionalization of rule of law, not segregation, but how long this is, precisely whose responsibility it is and at what point indigenous "spoilers" stop waiting out the security force and capitulate or reintegrate are all open questions.

As President Lincoln prepared for the war's aftermath, he asked German legal scholar Francis (Franz) Lieber in 1863 to attempt to codify the laws of war. This code of legal standards for imposing martial law was promulgated as General Orders No. 100.[14] The development of the

"Lieber Code" as a guideline for Reconstruction further muddied the waters, although it continues to be used today without most stabilization actors even being aware of it. [15]

The Lieber Code is in no way akin to a status of forces agreement (SOFA), as Lieber's premise was that the Code mandate the behavior of a "victorious army" toward an "invaded country" and how to manage such a postwar occupation. To this end, many of its provisions, while intended at the time simply to not destabilize during the acute risk period, would not be appropriate to stabilization and reconstruction operations today, particularly in their wording. Article 114 charmingly notes, "So sacred is the character of a flag of truce, and so necessary is its sacredness, that while its abuse is an especially heinous offense, great caution is requisite, on the other hand, in convicting the bearer of a flag of truce as a spy."[16]

Many of the Code's provisions, while intended at the time simply to not destabilize, would not be appropriate to stabilization and reconstruction operations today, particularly in their wording. The Code, did, however, provide some legal grounds for dealing with issues such as martial law, treatment of prisoners of war, enemies, victims, civilian property, public property, armistice, assassination and insurrection, and even treatment of women.[17]

The Code, did, however, provide some legal grounds for dealing with issues such as martial law, treatment of prisoners of war, enemies, victims, civilian property, public property, armistice, assassination and insurrection, and even treatment of women.[18] Article 102 is a particular example of gender equality: The law of war, like the criminal law regarding other offenses, makes no difference on account of the difference of sexes, concerning the spy, the war-traitor, or the war-rebel.[19]

The Code did not directly translate into doctrine nor provide specific instructions to commanders on how to govern their assigned territory nor state what the objectives of the occupation should be overall or in each sector or state. Later on, as President Johnson wavered and the Congress struggled to plow ahead with its agenda for the South, this lack of specific objectives and tasking resulted in a long, violent struggle to establish security, rule of law and stable governance over territory and inhabitants unwilling to yield to the occupiers' claim of authority. This sentiment still echoes today in some passionate assertions of states' autonomy and claims of overreach by "big government."

In order to establish safety, basic services and legal protections for emancipated slaves, the Freedman's Bureau was a significant institution run entirely by the U.S. Army in the occupied South. After President Johnson vetoed the original bill creating the Bureau as one that would exist permanently, Congress revised it in February 1866 to extend its mandate but make it provisional, expiring in July 1868, thus addressing concerns of opponents that it would permanently rewrite American legal code and of supporters who wanted to see former slaves protected longer-term.[20]

As stated by Colette Rausch of the U.S. Institute of Peace in 2006, "Without **public order**, the people will never build confidence in the public security system and will seek security from other entities like militias and warlords."[21] In concert with this statement, the Ku Klux Klan was formed in 1866, with members of this "social club" in Pulaski, Tennessee claiming it was only for defense of citizens in lands considered lawless by many. In fact, it became a collection of secret paramilitary organizations intent on bringing Democrats back to public office and reinstating white "power" across the South. By 1868 this "club" concept had spread into every Southern state.[22] Most local clubs or "cells" were not integrated with any other, all acted independently as a decentralized, clandestine militia within their own local territory, while sharing basic goals.

Such dispersion made these groups much more elusive as there was no large network to disrupt; of course, most locals knew who belonged and who the likely ringleaders were, but many white citizens supported them or, along with most black residents, were often too terrorized to speak up or try to stop their violent activities. The nebulous nature of these groups, united only in a chauvinistic ideology and a willingness to commit violence to support it, was akin to today's Islamic and other radical groups that are able to recruit activists via the internet without requiring an official joining process or membership status. Many of the contemporary terror attacks in the West have been perpetrated by these unofficial affiliates of a conceptualized larger movement. The Klan in all of its manifestations became the primary threat to public order and the rule of law in the Reconstruction period.

While we generally say we do not have interior ministry troops or a national police force, the U.S. Marshal Service is the enforcement arm of the judicial branch and has no state or local jurisdictional limitations on its activities. During Reconstruction, dozens of Marshals and their deputies wound up in Southern jails for attempting to enforce Federal laws in those territories, sometimes even on the evidence provided by those they were attempting to arrest. Others were arrested on murder charges after gunfights with criminals. Marshals were frequent targets of those resentful of Federal incursion on Southern soil as they attempted to enforce Federal laws in their local jurisdictions.[23]

The Posse Comitatus Act of 1878 was enacted as a result of the Compromise of 1876 and in defiance of the Army's efforts to maintain public order in the face of an increasingly assertive white population. The Act keeps the Army from conducting law enforcement activities on U.S. soil except in case of Federally declared martial law. This means the Army is prevented from policing activities including arrests, investigations and search and seizure. In other countries, the U.S. Army can only act in this capacity under the auspices of a Status of Forces Agreement (SOFA). In the case of occupation, the occupying power can act in all executive, legislative, and judicial capacities without a SOFA and generally according to continuing Lieber Code-established principles.[24]

Interpreting the Posse Comitatus Act became increasingly challenging into the twentieth century as it was "suspended" during World War II by the Secretary of War essentially "clearing the way for repeated violations in the years ahead." Throughout that century and into the next, violent eruptions over labor and race called for the military to augment law enforcement. Civil rights, antiwar demonstrations and labor disputes, some of which reached the level of riots, required the military to assist in maintaining civil order in numerous ways but there were no new convictions of military commanders for violating the Act.[25]

In the summer of 2020, the President has used Federal law enforcement officers, including prison guards in a number of cases to quell peaceful protests in June in Washington, DC's Lafayette Square[26] and in July in Portland, Oregon.[27] This incursion of Federal law enforcement, often under protest of local mayors and failing to identify themselves as Federal officers or wearing official insignia, has increased sharply with the administrations crackdowns on Black Lives Matter protesters, illegal immigrants and other popular backlashes related to the Covid-19 pandemic. The complexity of rules governing their jurisdictions and rules of engagement are likely to lead to further calls for increased regulation of Federal law enforcement.

During Reconstruction, the constant mustering out of local cavalries based on enlistment date left the least experienced and often black troops in place in South. This was seen as an affront to most Southern whites, who complained bitterly and offered up accusations against the actions of the soldiers. As Brigadier General Thomas H. Holden in North Carolina noted, "Now that the rebellion has been suppressed, it does seem to me that a great and magnanimous government like ours is not obliged to keep colored troops in our midst." With this in mind, General Thomas Ruger reposted many black troops to coastal and remote locales, while noting that, "the acts complained of, if committed by white troops, would probably not have been the cause of formal or persistent complaint."[28]

From the start of Reconstruction, white Southerners demanded black troops be relocated out of fear of their potential influence upon freed-people's new rights in terms of labor and political organizing. Through the 1860s, most black troops were shifted to the West and in September 1865, all black regiments were disbanded. By the end of Reconstruction, the only significant sites of black troops in the ex-rebel states were in west Texas, where they primarily fought Indians and protected white settlers.[29]

By the time the Republican effort to have President Andrew Johnson impeached was well underway, the Ku Klux Klan was already on the rise, aiming to undo the gains made since 1863 to unify the country and support the aspirations of Southern blacks. While the Klan initially claimed to form as a "law and order" augmentation where the Army was unable to maintain security, it became itself a key threat to public order. The Klan essentially became the local militia keeping the "peace" in a way that President Lincoln had not envisioned when he signed the Emancipation Proclamation or began to plan for postwar reconstruction.

In 1871, President Grant was able to get the Ku Klux Act passed, enabling the army to conduct targeted operations that greatly curtailed Klan activities. But in 1882, the Supreme Court declared the Klan Act unconstitutional. By the time the Klan Act was rescinded, its objectives had already been achieved through legislation in individual states and in the U.S. Senate, so the Klan ceased to be a factor as an organization while its political objectives became fully institutionalized into American legal code short-handed as "Jim Crow."

The Klan evaporated back into a few disparate social organizations as its chosen candidates increased their power at the ballot box and resurged during particularly violent periods, typically surrounding great economic dislocations and periods of black empowerment, as in the period following the 1896 *Plessy v. Ferguson* ruling upholding state "separate but equal" laws, the 1909 establishment of the National Association for the Advancement of Colored People, which used landmark court cases to attempt to reverse discriminatory laws and practices. Other periods when white supremacy resurged included the 1915 reestablishment of the Klan as the

"Invisible Empire," and again in the 1950s and 1960s in tandem with the black civil rights movement.[30] The late 2010s through the present period in mid-202 has seen increased attention on white police killing of black Americans, racially biased policing, fewer economic opportunities for people of color and removal of Confederate flags and monuments to slave owners and Confederate heroes referred to as the "Black Lives Matter" movement.[31]

These Klan reappearances also coincide with intentional spikes in Confederate commemorative building naming and monument erection, as shown in Fig. 3.3.[32] Today's debate about whether to retain such edifices is rarely elucidated by the fact that the majority of today's monuments in question were not placed there to honor Southern sacrifice directly following the war, but in later times as an intimidation effort and backlash against black assertion of the rights of citizenship as part of the perpetuation of the Lost Cause Myth and other white supremacy efforts to change the focus of Civil War history to the perceived injustices toward white Americans during Reconstruction.[33]

The Freedmen's Bureau endeavored to provide **Accountability to the Law and Access to Justice** to freedpeople and poor whites, but its resources were never sufficient, particularly when locally elected or appointed officials failed to administer justice without prejudice, thus removing accountability. Ironically, as W.E.B. Du Bois describes in his 1903 essay, *Of the Death of Freedom*, "The most perplexing and least successful part of the Bureau's work lay in the exercise of its judicial functions." He explains that the Bureau failed to maintain "a perfectly judicial attitude," by siding unwaveringly with black litigants against white plaintiffs. This did much to discredit all of the excellent work of the Bureau in the contemporary period and, which today has left its memory in obscurity, instead of recognized as one of the great moments of enacting freedom in American history.[34]

Sustainable Economy

As the Federal government sought funding to pay for the war and its aftermath, freedpeople endeavored to find their way in a new wage economy where their labor was undervalued.

The Federal government decreed that Confederate debt remain unpaid to prevent profit from secession and to differentiate the newly reunited states from their Confederate predecessors. This freed Southern states

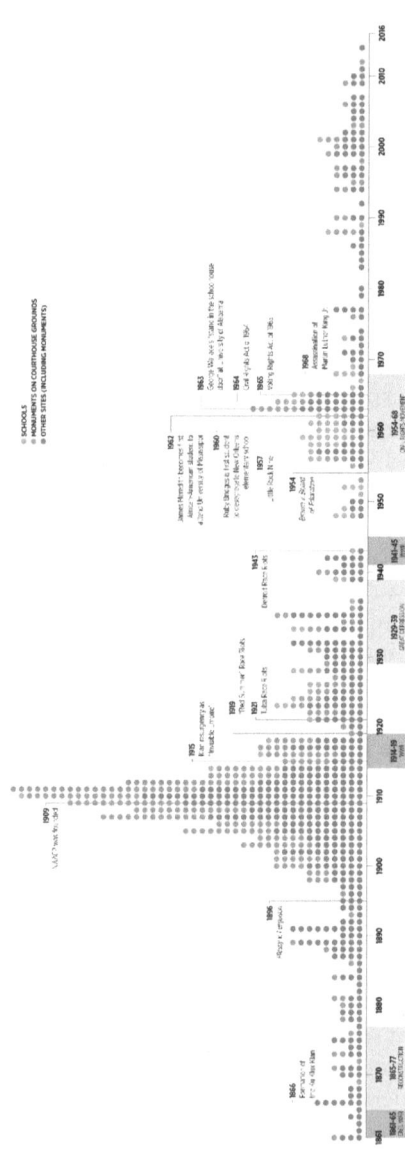

Fig. 3.3 Confederate monument establishment timeline (Courtesy of the Southern Poverty Law Center: https://www. splcenter.org/sites/default/files/com_whose_heritage_timeline_print.pdf)

Fig. 3.4 Sustainable economy

from war debt, but they still struggled to raise income after the devastation of the war and the huge cost of the loss of four million units of free labor (Fig. 3.4).

Northern states used excise and luxury taxes to generate revenues but such tariffs benefitted the North far more with its industrial products than they did the commodities-based economy of the South; tariffs for raw materials were 50 percent lower than those for manufactured goods.[35] To supplement its income, Southern states targeted freedmen with vagrancy and other laws to force them into a convict lease program that essentially continued the practice the Union had sought to destroy, vestiges of which remain within U.S. institutions to the present day.

Assessment

Opposition to emancipation had many manifestations from religious to cultural to political, but economics underpinned many of these. That is primarily due to the enormous economic benefit the Southern economy derived from slave labor. While the invention of the cotton gin increased productivity in the processing of cotton by an estimated 50 times, it's efficiency gains did not extend to the collection of cotton fiber from the fields where it grows. This was done entirely by hand and the expansion of the cotton gin only made cotton production more lucrative, leading to a desire for more planting land and with it, more slaves to grow it. In Northern states, industrial manufacturing was on the rise, providing economic opportunity for unskilled labor, including freed or escaped slaves.

The cultivation of cotton spread so much in the first half of the nineteenth century that by 1860, the majority of all the cotton produced globally came from the American South. The Confederacy tried to bring Great Britain to their fight, with its insatiable cotton appetite, by creating a faux cotton famine to inflate the price. This backfired and only enhanced the Brazilian, Egyptian, and Indian markets' competitiveness during the war and beyond. Therefore, slave labor or a facsimile was needed to ensure the price remained competitive so Southern plantation owners could regain their standard of living.

While Northern opponents of slavery characterized it as "weak, unprofitable and backward," Army officers entering stubborn slaveholding areas in 1865 learned that the institution was tenacious precisely because as the world's appetite for cotton continued to grow, slavery was "profitable, efficient, and powerful, embedded within a modernizing world, not a relic of the past."[36] As cotton spread into new states in the West, northern Republicans feared that more and more states would permit slaveholding, which could not only threaten the North's ability to maintain slave-free status, but low-skilled workers did not want to have to compete with no-cost labor for non-cotton jobs and were often the most vocal abolitionists.

A 1975 study calculated that the Civil War cost the United States $6.7 billion in 1861 dollars or two full years of gross domestic product (GDP), which were disproportionately shared. One per capita estimate suggests each Northerner lost $139, a little less than one year of GDP per person due to the war; each Southerner lost three times that amount in productivity, *added* to the cost of the human capital loss when slavery was abolished, which brings the per capita Southern economic burden closer to five years of GDP. [37] Of course, this burden was not shared equally among Southerners, as the vast majority were not slave- or even large-scale land-owners. Thus, the suffering of the elite plantation owners, who had held the majority of the power in the old South was multiplied exponentially, particularly among those who had also lost their lands after the war, along with their slaves.

After the 1866 Congressional elections, Northern "Radical" Republicans controlled the House and pledged to ensure citizenship, equality and economic freedom for blacks. The U.S. Army was employed to assist in developing a new economy based on freed labor.[38] This economic development element is critical to practical consideration of Reconstruction as a stabilization effort; agricultural economists have estimated that the slaveholding states also collectively lost $4 billion in human capital assets with the end of slavery.[39] This consideration is useful in understanding the economic catastrophe emancipation appeared to Southern state decision makers, without delving into the moral issue into which such discussions typically degenerate.

From the **Macroeconomic Stabilization** perspective and due to the naturally destructive nature of war, Southern infrastructure was in tatters. This was not only from the Union attrition strategy that destroyed nearly everything in General Sherman's wake across Georgia and the Carolinas, the best-known and extreme example, but not the only instance. Similar destruction of military targets, banks, hospitals, schools, businesses, farmland, livestock, railroad and telegraph lines, fences, irrigation systems, everything needed to sustain an army and a population willing to support it had been ruined across the South.

Even when the Union armies had not overtly destroyed infrastructure, but used it, as they had the Nashville & Chattanooga (N&C) Railroad to maintain Sherman's supplies, moving so many men and goods across the South by both sides resulted in grave disrepair. When the government returned the N&C to its private owners, they bemoaned the deplorable condition of the railroad due to the "ravages of war" and got to work trying to restore it "without tools, supplies or even…a chair or table or a sheet of paper" and more seriously, without "money or credit."[40]

All of this had to be rebuilt or improved upon to restore commercial activity across the entire country. The Federal Treasury Department issued bonds for reconstruction and set to work with the Army Corps of Engineers to build railroads and bridges, dredge the Mississippi and Ohio rivers and many other important projects crucial not only to resuscitating the nation's commerce, but supplying the starving and destitute people of the South. In addition to the loss of livestock and other foodstuffs, mills, and irrigation systems were destroyed, and dam destruction had flooded fields with saltwater with hundreds of more acres alternatively overgrown or scorched.

While the stories of outlaws like Jessie James have become legendary in American history, what is less understood is what lack of **Control over the Illicit Economy and Economic-Based Threats to Peace** arose during Reconstruction, such as moonshining, counterfeiting and cattle rustling, which offered employment to returning Southern veterans who may have lost their farms or lacked the requisite skills or connections to participate in the licit economy.

A dizzying array of illicit actors evolved in or descended upon the South as soon as the fighting ended. The terms bushwhackers, carpet-baggers, jayhawkers, and scalawags are vaguely familiar but their original meanings and differentiations across geographic areas are not necessarily clear today. These included outright criminals and others who skirted the edges of ethical propriety while turning a profit. All preyed upon the disarray and economic despair that resulted from the Southern defeat, but it is useful to understand the basic types of illicit actors that both drove and diminished Reconstruction's economic recovery.

The term "carpetbagger" remains in use today to connote low-class white men from the North who came South during reconstruction to buy up land on the cheap, speculate in cotton and generally make their fortunes from others' misfortune in the chaos and financial crises of the post-Civil War South. These people were said to be able to put everything they owned into a cheap bag made from old carpet and seem akin to today's "used car salesman" stereotype.

While there certainly were many of these, both of Northern and Southern persuasion, the great majority of those coming South were skilled professionals looking for work in the devastated South that was already short of a widespread educated class and also to serve as volunteers or in positions to assist poor whites and freedpeople in fields such as education, medicine, commerce, and law.[41]

Today's negative connotation of carpetbagger, likely came from the Southern tendency at the time to view *all* Northern migrants with extreme suspicion, especially when newcomers were intent on helping freedpeople or when their social standing exceeded that of the beholder. It is often used today to demean a politician with few ties to the place where he or she is running for office.

Another term still used vaguely today to connote a rascal, rogue or maybe some kind of pirate, is "scalawag." While this term was in use before the Civil War to indicate a farm animal with little value, like "runt," it evolved to mean a Southerner who supported Reconstruction. These were also called "copperheads," as in poisonous vipers. These were considered traitorous and even worse than the hated carpetbaggers and included farmers that did not have slaves and middle-class professionals who had remained Unionists throughout the war. Both of these words have become synonymous with swindlers of a very low type, but their true crime was in working with freedpeople to further the Republican Reconstruction agenda by banding together in voting blocs to support civil rights and economic development.[42]

More devastating to the efforts to rebuild and revitalize the South was violence and banditry, not just from the Klan's Night Riders but from other guerrilla bands mainly operating in border and other contested areas throughout and after the war. These criminals would raid military supply trains and even private farms and homes for food and loot. Many of these became the "outlaws" of legend after the war, like the James brothers, or migrated to Mexico and the western lands, graduating to bank and train robbery and horse rustling as well as preying upon the growing wagon trains heading west.[43] These kept the U.S. Marshals as well as the Army busy trying to maintain security, especially in the outlying areas distant from the Army posts and cities.

Economic stabilization requires effective **Fiscal Management**. While the Federal government did not want to take financial responsibility for actions that had been perpetrated against it, it also did not want to suggest that the newly welcomed states bore any relation or resemblance to those that had seceded. In order to delegitimize the Confederacy, Section 4 of the 14th Amendment states that "neither the United States nor any State shall assume or pay any debt or obligation incurred in aid of insurrection or rebellion against the United States, or any claim for the loss or emancipation of any slave; but all such debts, obligations and claims shall be held illegal and void."[44] This repudiation of Confederate debt provided early relief to the states as they attempted to rejoin the Union and begin to rebuild. However, it also proved disastrous to those who had invested in the rebellion, who were often pauperized after the war debts were repudiated.

Unfortunately proving the "political primacy" aspect of stabilization, even economic policy decisions were made through a political lens and often at the expense of growth. The Northern population had endured income and excise taxes as burdens of war but expected them to be lifted once the fighting stopped. Reconstruction would be expensive, so the Federal government tried to find ways to raise revenues in the short-term, including maintaining sales taxes on alcohol, tobacco, and some luxury goods, which comprised 50 percent of government revenues by the 1880s.

Tariffs were a complex issue, which Southern apologists consistently maintain were the "real" reason for secession, and not slavery. This remained a point of contention after the war, as manufactured goods, mainly produced in the North had tariffs as high as 100 percent and raw material tariffs, such as on cotton and sugar produced in the South, were half as high, leaving these growers less protected from foreign competition.[45]

In addition to the obvious costs of war, in March 1865, an additional economic crisis left the Union government with larger than anticipated war debt that contributed to the push for demobilization of the standing army, thus reducing the number of total troops available for sustaining effective stabilization in the South. As many others have meticulously described its cause and effects, here is a very brief discussion of the crisis as it affected the postwar **Market Economy Sustainability** process.

By the end, the estimated cost of the war was up to nearly $6.7 billion (just over $105 billion in 2020),[46] which the Federal government mainly financed by selling war bonds, and which ultimately linked government borrowing practice to the value of gold, as interest payments were promised in this medium rather than in volatile paper currency. However, as Gregory Downs explained, "If gold rose too high, the government incurred huge interest payments. If it dropped too low, the loans and bonds lost their financial appeal."[47] At that time, the anticipated Union victory caused the gold price to go down, which greatly reduced public demand for government bonds and other lending instruments, causing another drop in demand for "railway shares, petroleum, pork, tobacco, and cotton by more than 20 percent."[48]

Through some clever schemes, the government managed to avert crisis, but scarred Treasury Secretary Hugh McCullough then attempted to sharply curtail expenditures. Therefore, despite the ongoing conflict in North Carolina, Texas and Virginia in Spring 1865, as well as other pockets of the massive, restive Southern territory, planning focused on mustering out troops as quickly as possible. The ultimate plan called for the immediate halt to the draft and reducing procurement and personnel to the tune of over 800,000 Union troops demobilized in time for Thanksgiving 1865, leaving insufficient resources to provide effective security.[49]

Most crucially, the War Department sought revenues by selling horses and demobilized the cavalry too quickly, more than decimating the number of these soldiers who were able to cover 30–50 miles in a single day to a woefully small series of infantry forces who could only cover up to 5–10 miles in a day and only with rare efficient railway connections. The Treasury made $7.5 million by Autumn 1865 selling 128,000 horses and mules and removing cavalry forces entirely from some Southern states with numbers dropping in still-restive states by astounding rates, such as 10,000 in North Carolina at end-war to 854 by midsummer.[50]

This profit and the reduction in cost of maintaining the quadrupeds' riders likely alleviated some of Secretary McCullough's initial worries after the March crisis that demobilizing troops, some of which were owed up to six months of back pay, would cost $50 million in the short-term.[51] Akin to the drawdown of combat operations in Afghanistan and Iraq, the plan was to increase the number and complexity of the tasks the Army was to undertake during stabilization while drastically cutting its resources, with unsurprisingly similar results. Over and over in U.S. history, as Downs sagely opined, "war proved more popular than occupation."[52]

There was a critical need for **Employment Generation** in the South, particularly for 4 million new free labor workers needing to be placed in paying jobs, as well as for poor whites and others who had lost their farms or main breadwinners through death or injury during the war.

Although many freedmen moved north or west for work, those that remained were not all turned out of the plantations where they had worked. Some rented land from planters and were often lent seed, tools and other equipment on credit to be paid back, along with an agreed portion of the next harvest in exchange for the use of the land. This appeared to be a generous arrangement for freedpeople and some land-less whites who became such "sharecroppers," but farming is a capricious master and a good harvest is not guaranteed. Many sharecroppers ended up deeply indebted to their landlords and although family members were no longer sold off and sharecroppers were ostensibly not allowed to be beaten, for many, not much seemed changed after emancipation.

In some areas, cotton and sugarcane did not dominate the local economy so the farms were smaller, had employed fewer or no slaves and thus, those freedpeople had more diverse skills than those in strictly agricultural regions. In more arid or hilly regions, Reconstruction-era employment could be found in mining, timber, milling, railroads, and carpentry. Obtaining wages allowed freedmen to purchase land that had been abandoned or from poor whites to augment their income with farming or livestock production. Many whites continued working in agriculture but also found employment in mining, timber and in the growing textile and tobacco factories.[53]

While the various outlaw types caused economic setbacks on a small scale, the more insidious crimes of the Reconstruction era were those legally perpetrated by the states themselves. Although the Emancipation Proclamation ostensibly freed slaves in 1863 and General Order No. 3 was issued in 1866 with a provision that ensured freedpeople were not to be charged with "offenses for which white persons are not prosecuted or punished in the same manner or degree,"[54] the Southern economy was so dependent upon slave labor that plantation owners and others found creative ways to keep as many freedpeople in servitude as "legally" possible.

Although the practice existed before the Civil War, in the Reconstruction period, a number of vagrancy laws were enacted and those on the books were more rigorously enforced in many Southern states under which the unemployed, typically blacks or unskilled, poor whites, could be imprisoned when unable to pay fines. Eventually these hapless convicts became part of a complex web of privately leased labor by which they became almost property of the state able to be rented out as labor with a percentage of the wage and housing cost returned to the state as income. In Alabama alone, the profit from this convict leasing program was estimated at 10 percent of the total state revenue in 1883 and by 1898, it had grown to 73 percent.[55]

As a 2015 Justice Department report investigation of the Ferguson, Missouri police department emphasizes, a similar revenue generating practice was at the heart of the 2014 riots, where black residents were disproportionately targeted for various traffic and other minor violations from which fines and court fees mounted, resulting in many ending up essentially in debtors' prison, often for years. These charges include "Manner of Walking" (95 percent) and "Failure to Obey" (89 percent) charges.[56]

The conditions under which nineteenth-century convicts were housed were often worse than some slave conditions had been. Death rates among such convicts have been estimated at 10 times higher than for other freed blacks with 25 percent of all black leased convicts dying in 1873. They were used in the cotton and sugar industries, but also in mining, logging and building railroads. Some historians have suggested that an additional incentive for convict leasing was to prevent labor union development and expansion.[57]

The practice continued legally until 1928,[58] but the argument can easily be made that with the high proportion of black inmates in for-profit prisons and escalating traffic and other fines resulting in increased incarceration for the poor, the practice is continued today in myriad other ways. President Franklin Delano Roosevelt halted the practice in Federal law on December 11, 1941[59] but vestiges of it remained, including pressing minorities, poor whites and women into chain gangs leased to private firms until 1994 in Alabama.[60]

With the expansion of prison privatization from the 1980s through the present period, incarceration rates need to be kept at a certain level to ensure the profit margin, which has resulted in firms that administer prisons and even guard unions lobbying for state and federal legislation that lengthens sentences and increases penalties. These prisons are also often located in rural areas and provide critical employment opportunities for which locals also support such legislation as crime prevention, without even realizing the negative effect on the inmates or the hefty margins for the corporations they serve.[61]

By 2001, this process led the American Civil Liberties Union (ACLU) to declare the "War on Drugs," to be the "new Jim Crow," noting that the "number of black men in prison (792,000) had already equaled the number of men enslaved in 1820." Americans convicted of felony offences permanently lose the right to vote and, as felonies are defined by state, in Florida, for instance, minor drug crimes count as felonies.[62] A 2016 report by *The Sentencing Project* determined that more than 7.4 percent of the U.S. adult black population is disenfranchised in this way but only 1.8 percent of the non-black U.S. population. The highest rates, over 20 percent, are in the states of Florida, Kentucky, Tennessee, and Virginia.[63]

In November 2018, Florida voters approved an amendment to the state Constitution that would restore convicted felons' right to vote once they have served their requisite prison time. In May 2019, Conservative lawmakers, recognizing the large proportion of reinstated felons are black and poor, passed another law requiring them to repay all fines and court costs before they can register to vote, a move guaranteed to limit the number of released convicts who will be able to participate in elections. This new law was signed by Florida Governor Ron DeSantis in June 2019.

STABLE GOVERNANCE

Reconstruction is often described as having had two phases so distinct, they are called "the first" and "the second" Reconstruction periods. Much of this has to do with the inhabitant of the White House during the two periods. The first was dominated by the Legislature after President Lincoln's assassination and the second, disorganized phase, under President Johnson. President Lincoln had wisely begun planning for postwar stabilization in 1863, developing the Ten Percent Plan and other mechanisms for reconciliation and infrastructure development projects to modernize the economy of the South, such as railroad and other infrastructure expansion, to blunt its dependence on cotton once slavery was abolished. That planning changed abruptly with Lincoln's death (Fig. 3.5).

The Army endeavored to provide **Essential Services** in early Reconstruction, primary of which was security, without which no other services can be considered. In addition to this basic security, the Army often served as a transitional public security (TPS) force. Army commanders tried to work through local officials but often had to replace mayors, magistrates, judges, and police leaders for prejudicial behavior toward freedpeople. The Army managed to keep the peace and enable limited civil rights for freedpeople where access was possible, educating them about these rights and handling grievances and managing disputes.

Fig. 3.5 Stable governance

STABLE GOVERNANCE

• Provision of Essential Services
• Stewardship of State Resources
• Political Moderation and Accountability
• Civic Participation and Empowerment

As with all services provided by the Army, as it drew down in the South, areas of accessibility and, thus, civil rights protections and health and education services for freedpeople shrank to the areas within five miles of Army posts and in larger cities just as states began to reassert their influence through restoration of prejudicial officials and legislation. Volunteers in the Freedmen's Bureau strove to maintain services, but by 1872, the organization had been disbanded and any remaining activities were conducted under the War Department.[64]

One of the major oversights in the effort to establish stable and sustainable governance was the military commanders leading occupation seeing the elite slave- and plantation-holders as traitors to be left out of the process. One of the critical approaches to the condition of *Political Moderation and Accountability* is to *Build broad-based consensus on the country's political future through inclusive and participatory national constituting processes*. The *Guiding Principles* note "Elites play a particularly important and powerful role in this form of political settlement, and their buy-in and support is critical. The inclusion of women, minorities, and non-state traditional institutions is fundamental to the success of national constituting processes."[65] These commanders left out the elites and the women to their and posterity's peril.

Assessment

President Andrew Johnson, a Southern Democrat, understood that the war had been fought over slavery and that the abolitionists' side had won, therefore emancipation was a foregone conclusion. He was the only Southern senator to declare his loyalty to the Union after his own state seceded, making him a hero in the North but reviled in the South and especially in his own state. Although Johnson himself owned some slaves, but his relatively humble roots did not make him sympathetic to the white planter class and after secession was widely quoted as saying, "Treason must be made odious and traitors punished."

Johnson also did not initially support emancipation and lobbied for an exception to it for Tennessee as that state's military governor. In 1863 when a pro-slavery civilian governor was elected, Lincoln instructed Johnson to ignore the vote, ensure emancipation became part of Tennessee's constitution and only allow pro-Union candidates to participate in subsequent elections. By forcing those who wished to be enfranchised to take a loyalty oath to the Union and then wait six months, he ensured that "radical" Republicans took the next elections.[66] Neither of these actions endeared him to the Southern elites.

Johnson did not favor granting full civil rights to freed blacks, which caused conflict with Northern Republicans. He also granted clemency to large numbers of Confederate leaders, including Alexander Stephens, who had served as Jefferson Davis' Vice President, further angering Unionists. He vetoed both the Civil Rights Act of 1864, overturned by the legislature, and the first Freedmen's Bureau bill, finally signing a second one with an expiration date.[67]

When Congress passed the Tenure of Office Act, which gave the Senate, not the President, the ability to remove Federal officials, after he had himself removed 1600 postmasters from office, thus threatening the system of patronage that allowed leaders across the government to ensure their own candidates got lucrative political appointments, it was clear he had lost favor with both parties. Johnson claimed his positions were based on a need for national reconciliation, but he was ultimately saved from impeachment by only one vote, illustrating the continuing divide in the nation that had a long healing process yet to undergo.[68]

The 1868 election of General Ulysses S. Grant as President was a watershed moment in the process of Reconstruction. General Grant had won fame as the supreme commander of Northern forces in the war and as President, would champion full rights for freedmen. He was also privileged to preside over a Republican-majority Congress, with which he worked well to implement the Ku Klux Act passed on April 20, 1871 that stated "insurgents were in rebellion against the authority of the United States," thus permitting Army commanders deployed around the South to conduct operations against the Klan and its various adherent groups including declaring martial law and suspending *habeas corpus*.[69]

While these efforts were effective in quelling the worst of the violence, by the 1876 Presidential election, as all Southern states had rejoined the Union, the Democrats had regained the House of Representatives and the Northern Republicans had become weary of Reconstruction with the war far behind and the abolition of slavery appearing to have been completed. The North erroneously believed reunion meant reconciliation, while violence ramped up again in an effort to intimidate freedmen and their supporters away from the polls. Although most accounts indicate that the Army did a good job keeping them away from the actual polling stations, the intimidation campaign still had the intended effect, ensuring Democratic majorities were elected across the South.[70]

The 1876 presidential election was too close to call and a commission was created with leaders from all three branches of government deliberating on what to do. In what became known as the Compromise of 1877, and often viewed as one of the shadiest of backroom deals in U.S. history, Republican candidate Rutherford B. Hayes agreed to send the Federal troops home if his election bid against Democrat Samuel J. Tilden went uncontested, thus ending Reconstruction.

In terms of *Stewardship of State Resources*, the Federal civil service expanded from 5837 in 1861 to 15,344 (excluding 30,000 postal workers) by 1871,[71] enhancing efficiencies recognized during the war. As Southern states rejoined the Union, their Democratic state legislatures rightly claimed these institutions were rife with cronyism and sought to reduce this and counterbalance the Republican Federal legislature and curb the Executive Branch's power. Major abuses also took place in Southern state legislatures once home rule was reinstated. Public outcry on both sides eventually restructured the system to one intentionally based more on merit than social networks, through the Civil Service Act of 1883, with this process continually improving ever after.[72]

One of the most forward-thinking initiatives of this period was the passage of the Morrill Land Grant Act of 1862, which bestowed a parcel of land upon each state remaining in the Union to develop agricultural and mechanical training colleges.[73] In 1890, the program was expanded to the remaining states, who had previously refused to use land grants to support education for freedpeople. These became predominantly black colleges and universities, called "1890 Institutions,"[74] providing needed higher education opportunities and producing the engineers and other technically skilled labor needed to continue the rapid pace of infrastructure development. One of these now called "Historically Black Colleges and Universities (HBCUs)" is Howard University in Washington, DC, founded by General Oliver Otis Howard, the Freedmen's Bureau Commissioner. Howard was known for his Christian piety and the school was originally founded for African American preachers to study in 1866, but later expanded its offerings to include liberal arts, becoming a university and finally adding a school of law.[75]

The Federal government provided supplies to starving Southerners through the Army, but this was made difficult by the destruction to critical infrastructure such as railroads and bridges, which the Army was also tasked to repair and rebuild as quickly as possible. In this effort, the Army was eventually assisted by the New York-based Southern Famine

Relief Commission (SFRC), founded in 1867 by several officials that had worked extensively with the Sanitary Commission.[76]

Federal funding for rebuilding and expanding infrastructure in the South was notoriously mishandled, with nearly half by many estimates going to other projects or individuals' pockets. In one famous example, an opera house and hotel were built with funds intended to repair and expand the Alabama and Chattanooga Railroad.[77] Despite such a notable example, the amount and diversity of infrastructure and rebuilding that took place during the Reconstruction period and the contemporaneous migration West, particularly under the direction of the Army Corps of Engineers, is simply staggering and led directly to the enormous economic growth that took place into the next century.[78]

In terms of **Political Moderation and Accountability**, the U.S. Army had been deployed to the South to ensure Republican state governments survived, which disintegrated when the Army was removed as part of the Compromise of 1877. As in today's conflicts, it is difficult to declare stabilization "complete," as competing factions and territories are likely to prove more or less amenable to acquiescing to or participating in stabilization measures, whether they are imposed by an external military force or by a host nation perceived by sufficient numbers to be hostile.

While providing basic security is critical to stabilization, once people start to resume their normal activities, they begin to demand many other services, such as free flow of goods to the markets, reopening of schools and employment opportunities, forgetting how only a short time ago, their main preoccupation staying alive. This was a poignant observation in General Raymond Odierno's political advisor Emma Sky's book on the shortcomings of stabilization in Iraq, *The Unraveling*.[79]

Even in the United States during Reconstruction, it was clear that alternative governance structures that threatened the sovereign's ability to impose order and conduct stabilization arose for various reasons, providing services or inspiration to some elements of the population. This makes identifying and applying appropriate metrics to stabilization progress even more difficult as both the operations and their success metrics cannot often be applied evenly.

Alternative governance still exists with Federal government approval in parts of Appalachia where mountains and hidden valleys made already independent-minded Scots-Irish families, like the fabled Hatfields and McCoys, both deeply patriotic yet intensely suspicious of both the Federal and state government. To make matters worse in that region, the coal industry with government complicity created towns that slavishly held generations captive to this single industry without concern for citizens' welfare or its impact on the local environment, only focused on the productivity of the mine.

Even as late as 2010, when an explosion claimed the lives of 29 miners, the company still continued to operate with impunity, blaming the employees for not ensuring safety standards were maintained as coal productivity quotas steadily increased.[80]

This kind of patriotism coupled with mistrust of government arose sharply during Reconstruction, as the Union forces were literally occupying the South. The failure to include women in any post-conflict peacebuilding efforts ensured that there may have been reunion, but certainly not reconciliation. Women were arguably the most powerful actors to ensure the Southern myth of the Lost Cause is perpetuated even today.

At the end of the war, women took the lead in finding, identifying and transporting dead soldiers to sanctioned Confederate cemeteries that they also established and tended. The Federal government created national cemeteries, beginning with Arlington on General Robert E. Lee's Virginia homestead as an intended insult, but they initially did not create or share cemeteries with Confederate fallen. This task was taken up with gusto mainly by Confederate women and private donors they solicited. Throughout the postwar period, Confederate women were at the forefront of decorating graves and hosting various memorial day parades to keep the memory of their valiant men alive.[81]

Rather than waning with the passage of time since the war, Southern women increased their efforts to ensure their history would be the one told, not just in the South, but across the whole country. By the 1880s, these ladies' memorial societies with various names and iterations would move beyond honoring the dead where they lay to enter many other spheres including children's education. The monitored textbooks to ensure the words "rebel," "traitor," and any other pejorative terms for the Confederacy would not be used in classrooms. They perpetuated the Lost Cause through summer camps and other "educational" arenas and were they great keepers of the myth within families and communities. One author posits that "under their watch, the Lost Cause [began] to eclipse the Union Cause":

> White Southerners insisted they had fought in defense of hearth and home rather than a war for some abstract principle like the Union. They had waged a defensive war against those who had invaded their land – threatened their homes, freed their slaves and harassed their women and children.[82]

This focus on home and hearth fed naturally into the additional myth of the rapacious black man bent on raping and deflowering white women that would prove pervasive in the culture into the twenty-first century. This fear of polluting the blood and defiling women became a dog whistle for political violence, including lynching and other extrajudicial violence, police brutality, and other forms of discrimination to the present day whenever black men attempted to assert their civil rights, guaranteed in the 14th Amendment in 1868.

As efforts were made across the South to ensure **Civic Participation and Empowerment**, Democrats who opposed Reconstruction regained a majority in the House of Representatives in 1874 while Northern voters felt the Civil War and slavery were long over and fatigue for Reconstruction activities had set in. The role of the Army became ineffectual as support for Reconstruction waned ten years after the cessation of large-scale violence, while Southern opposition gained ground as the North failed to ensure the rights and freedoms granted were fully implemented and institutionalized and access to opportunity was granted across the entire population.

If there was this level of disinterest in effectively conducting stabilization operations in one's own country, it is understandable that the international community has difficulty maintaining the commitment in distant places like Somalia or South Sudan, where there is no direct national interest. There is much discussion today in stabilization circles about **Political Primacy**, meaning that all stabilization efforts must be focused on establishing and preserving positive and sustainable political outcomes, especially post-conflict. The politics of the post-Civil War period provide a useful lens through which to view the efficacy of early Reconstruction.

As noted above, the level of white supremacist influence and activity varied across the South, due to local conditions. In Virginia, for instance, where the Republican Party did not easily become firmly entrenched, the Klan was less active as its goals could be achieved through the political process.

In Georgia and Louisiana, however, the violence was frequent, brutal, and widespread. This directly correlated to the results of the presidential election of 1868 provided in Fig. 3.6 in which Republican Ulysses S. Grant, won the vote in every Southern state except Georgia and Louisiana. Somewhat surprisingly, in several Northern and border states the popular vote also did not go to the Republican. As Grant's opponent, Horatio Seymour was from New York, thus his wins in that state and neighboring New Jersey are less surprising.

Delaware, Kentucky and Maryland, all border states, also did not have majority support for Grant, perhaps because Kentucky and Maryland each had counties where the population was majority black. Although the 15th Amendment granting blacks the right to vote had not yet been ratified in these states, the white electorate, although mainly not in favor of slavery, was likely to have felt more threatened by the possibility that the black vote could have enough weight to determine the outcome of future elections.

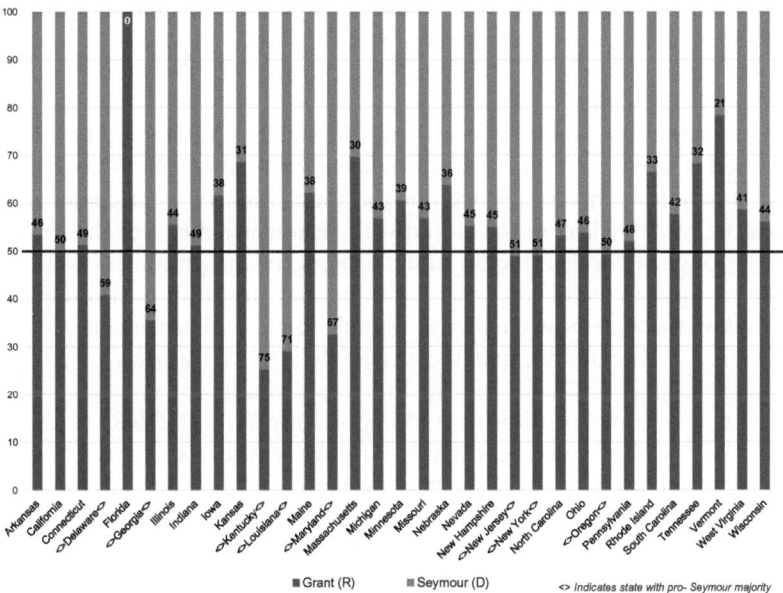

Fig. 3.6 Popular vote results for 1868 presidential election

Delaware seems an odd outlier today but it had been a slave state and while it had no majority black counties, it, along with Kentucky, rejected the 15th Amendment, that would grant all men the right to vote, and never actually ratified it until 1901.[83] This is particularly interesting as Delaware, along with North Carolina, Pennsylvania and Tennessee, had earlier permitted "free men-of-color" to vote, likely due to their small numbers at the time and therefore, marginal effect on the overall vote.[84]

In Delaware, opposition to slavery led white landowners to grant freedom to so many slaves in the early 1800s that fully half the black population consisted of freed slaves or 13.5 percent of the entire state population of 1800. By 1810, Delaware had a larger freed population than its neighbors at 76 percent over New York's 63 percent or New Jersey's 42 percent.

By 1850, there were no slaves left in the state and by the start of the Civil War, free blacks represented 18 percent of the total state population, numbering nearly 20,000. This increasing size of freed blacks in the state made Delaware's legislature nervous about them becoming a serious voting bloc, which resulted in increasingly draconian laws hampering political and economic participation by free blacks, a practice that mirrors precisely what happened across the South throughout Reconstruction on a much larger scale.[85]

Sustainability is also a key theme in stabilization today, borne of the realization that there may be relative calm immediately following the cessation of major violence but the foundation of what comes next must be carefully laid and nurtured with an understanding of the drivers of the original conflict and an understanding of the grievances of those who engaged in violence. A theme throughout this study is also the lack of follow-through post-conflict in completing the disarmament, demobilization, reintegration, and reconciliation (DDRR) aspects of ending conflict.

The most complex, difficult, and important parts of DDRR are always the "Rs," which must be extended to the entire society, often over decades. In consideration of the "political primacy" principle, the political environment can change dramatically post-conflict in a short time, particularly when the authority charged with DDRR is not given the resources to do so effectively and then is even forced to abdicate its role and leave the area entirely, due to a change in resourcing or political climate.

Planners and analysts frequently point to stabilization "successes" in Germany and Japan, but although the occupations officially ended in

1952, U.S. commitment to these societies and their stabilization has lasted over 75 years since the violence ended. The United States only maintained its Army in the South for 12 years after the end of Civil War combat in 1865, thus failing to fully reintegrate and reconcile the elite power into the reestablished Union and its new policies and social changes.

The two maps comprising Fig. 3.7 indicate how the situation changed from 1868 to 1880 in the South. The colors indicate **blue** for Democrats and **red** for Republicans with the depth of the colors showing the general percentage votes in each precinct. The paler the color, the closer the vote to 50 percent the darker the color the more unanimity in voting with the darkest colors indicating up to 100 percent of voters electing a single party. These maps show how the Democratic Party regained power after the Army was forced to end occupation following the Compromise of 1877. Once all the states were permitted to rejoin the Union with full (white male) citizenship rights restored and legislatures reinstated, the Republican agenda was essentially halted and the rights of citizenship for blacks prohibited for a century.[86]

Fig. 3.7 Comparison of 1868 and 1880 voting patterns in Southern U.S. States

Social Well-Being Introduction

As a singular stabilization effort to provide social well-being, the Freedmen's Bureau would provide a perfect case by itself on which to apply the *Strategic Framework*; however, the findings would support what is already well-understood about stabilization and reconstruction then and efforts since: the Army is always tasked with too many responsibilities, many of which are outside its core competency areas and it is rarely appropriately resourced nor kept in place long enough to achieve its strategic objectives (Fig. 3.8).

Through the Army, the Freedmen's Bureau endeavored to provide basic food, water, and shelter in displaced person camps that sprang up around the Army posts and on redistributed confiscated and abandoned lands. This was a difficult task, due to famine conditions in the South immediately postwar and the dismal state of infrastructure to produce and transport food and supplies. Conditions varied among the camps, due to the need, availability of resources, and the skills of the officers administering them.

As Southern states reentered the Union and regained home rule, by 1876, U.S. voters, particularly in the North, where they could not see the imposition of the Black Codes, which initiated the century of "Jim Crow," determined that Reconstruction was no longer necessary now the war was long over, and the slaves were free. With this failure to recognize the bifurcation of American society directly related to the failure to

Fig. 3.8 Social well-being conditions

SOCIAL WELL-BEING
- Access to and Delivery of Basic Needs Services
- Access to and Delivery of Education
- Return and Resettlement of Internally Displaced Persons
- Social Peconstruction

sustain Reconstruction, subjugation of black Americans has continued in different forms with different names to the present day as reflected in the declarations of the *Black Lives Matter* movement calling for equality and police and judicial reform. Racial divisions that have remained all but invisible to many white citizens since the Civil War are increasingly visible in the era of ubiquitous breaking news.[87]

ASSESSMENT

Wisely anticipating the scale of demand for **Access to and Delivery of Basic Needs Services** for the newly freed and the displaced poor, Congress passed a bill creating the Freedmen's Bureau in March 1865. General O.O. Howard was chosen as Bureau Commissioner, a man called by his many admirers the "Christian soldier" who, by most contemporary and historical accounts was the best man for an impossible job. At one point, General Sherman himself advised that it would not be in Howard's power "to fulfill one tenth of the expectations of those who framed the Bureau...I fear you have Hercules' task."[88]

Du Bois agreed with Sherman's assessment of the task as "a curious mess he [Howard] looked upon: little despotisms, communistic experiments, slavery, peonage, business speculations, organized charity, unorganized alms-giving - all reeling on under the guise of helping the freedmen, and all enshrined in the smoke and blood of war and the cursing and silence of angry men." He further described General Howard as, "An honest man, with too much faith in human nature."[89] As many would agree more than a century later, he was the right man for a thankless job.

General Howard divided the Southern territory into 10 Bureau Districts, each led by an Assistant Commissioner. Although the Bureau's efforts have been discussed throughout this study as the services it provided span the *Strategic Framework*, the critical point is that it was the Army that was charged by Congress to conduct myriad tasks related to Social Well-Being including providing basic access to food, shelter, education, medical care and legal services. As Gregory Downs described, "Where black troops were stationed, freedpeople flocked to them, and their barracks became centers of educational, religious, and political life. The enormity of the problem is eloquently captured with W.E.B. Du Bois' lyrical 1903 description of these desperate freedpeople as a "dark human mass that clung like remorse on the rear of these swift (troop) columns, swelling at times to half their size, almost engulfing and choking them."[90]

Before the war, the United States had primitive **health services**, especially in rural areas where country doctors would often not be aware of new medicines and procedures that slowly drifted across the Atlantic from Europe. During the war, the lack of understanding of infection, the overwhelming need for care, due to the extent of the violence, malnutrition and fatigue weakening soldiers as well as exhausting doctors, resulted in more than twice as many disease-caused deaths as those from battle wounds during and immediately after the Civil War.[91]

At the start of the war, the Union Army medical division had fewer than 100 surgeons. On the battlefield, there was no ambulance system for efficiently transporting wounded soldiers to hospitals or even field clinics. Thousands of soldiers lying in the open air while their wounds fester are common cinema images and a tour guide in Gettysburg describes the limbs piling up as high as the second-story windows of one house converted to a hospital where body parts were simply tossed out after amputations.[92]

The United States Sanitary Commission was established in 1861 in the North and began addressing many of these issues, organizing volunteers, such as the legion of female nurses that rushed to help the war effort, but this was only a starting point as the need was so great and the war so widespread and violent. The lack of consistent and modern care as well as the sudden engagement of thousands of men from across the country sharing confined spaces with open wounds and weakened health were easy prey for diseases such as malaria, smallpox, and tuberculosis, which men from various regions had never encountered and to which they had no immunity.

Typhoid was also a major concern with the National Institutes of Health (NIH) today estimating 75,000 deaths. Dysentery caused almost 100,000 deaths distributed nearly evenly on both sides. After the war, cholera raged across the South, as destroyed water and sewer systems became filled with detritus including livestock and human remains quickly decomposing in hot temperatures with the lack of understanding of these diseases and how to control them and adding to the infrastructure and human toll during early Reconstruction.[93]

The Freedmen's Bureau struggled to provide basic education and skills training to the many it found in its charge, but there was no standard consistent model of what and how to teach. Most freedpeople and poor whites had never before had **access to** even rudimentary **education**. Fear among slaveholders was pervasive that allowing slaves to become literate would lead to subversion, so it was generally prohibited. The proportion of young people of all races enrolled in school remained relatively low in the last half of the nineteenth century. Although enrollment rates fluctuated, roughly half of all 5- to 19-year-olds were enrolled in 1860. Rates for males and females were roughly similar throughout the period, but rates for blacks were much lower than for whites (59.6 white to 1.8 black per 100).[94]

The Federal government clearly expressed its support for universal public education by floating generous bonds and encouraging volunteer teachers to head south, but school funds were often diverted for exorbitant school and district administrator salaries, which were frequently paid even when schools were not being built or operated. W.E.B. DuBois estimated that freedpeople also contributed $785,700 (over $14 million in 2017 dollars) to Southern schools.[95]

Each state constitution lauded the importance of public education, but the resulting policies rarely met the rhetoric. Despite the difficulties, the number of school buildings, teachers and students enrolled across the South more than doubled from 1860 to 1871. Following the Civil War, enrollment rates of black students rose rapidly from 10 percent in 1870 to 34 percent in 1880.

Due to white fears of integrating the races in the classroom, in some locales, schools were vandalized or burned down as quickly as they could be built, and teachers were threatened with violence and even attacked. As a result, the cost of maintaining a "dual school" system for black and white children, particularly in sparsely populated rural regions, placed an unsustainable burden on available resources, especially as they were already halved by diversion and embezzlement by some estimates. From 1880 to 1890, there was essentially no change in the enrollment rates for blacks and the rate for whites actually fell. All of this led ultimately to extremely low education rates for both in many places, far lower for most black children.[96]

The literacy disparity shown in Fig. 3.9[97] indicates that in 1880, nearly 90 percent of the white population was literate and only 30 percent of the black population. This figure has clearly improved in the intervening century but according to a 2013 U.S. Department of Education study, of the 14 percent of native born Americans who are unable to read proficiently, only 9 percent are white and 24 percent black, indicating a continuing lag in the education of black citizens.[98]

Illiterate Percent of Population[147]					
Year	Total	White			Black and Other
		Total	Native	Foreign-born	
1870	20.0	11.5	—	—	79.9
1880	17.0	9.4	8.7	12.0	70.0

Fig. 3.9 Illiterate percent of U.S. population, 1870–1880 (U.S. Department of Commerce, Bureau of the Census, Historical Statistics of the United States, Colonial Times to 1970; and current population reports, series P-23, Ancestry and Language in the United States: November 1979 [This table was prepared in September 1992], available from https://nces.ed.gov/naal/lit_history.asp. Accessed on July 7, 2017)

The Freedmen's Bureau was intended to also manage abandoned land and property seized from leading whites, such as Robert E. Lee and Jefferson Davis or steadfastly unrepentant Confederates. Some of this property was subdivided and distributed to freedpeople and poor whites, but much of it would be retaken once the constituent state returned to home rule and Black Codes were instituted, denying blacks the right to own property.

In one prominent example, in January 1865, after his devastating march, General Sherman issued *Special Field Order No. 15*, confiscating 400,000 acres of abandoned coastal land from South Carolina to Florida. This land was sold to freedpeople in 40-acre plots and a later order offered a mule to each farmer, this is remembered historically as the "40 acres and a mule" promise.[99]

The Freedman's Bureau was formed the following March and mandated to distribute this land. Sherman also left Inspector General Rufus Sexton to attempt to administer and protect these lands. Unfortunately, this effort at early restitution and resettlement of freedpeople, which would have provided an estimated 40,000 of them with sanctuary and economic opportunity, was rescinded by Fall 1865 by President Johnson, who returned most of the land back to its original owners in capitulation to white landowners, thus exacerbating the continuing practice of rescinding civil and economic rights guarantees for black Americans that continues to echo today.[100]

Refugees from other countries were not a primary issue post-Civil War but **Return or Resettlement of Internally Displaced Persons (IDPs)** within their own country was one of these Herculean efforts with which the Freedmen's Bureau was tasked. While the Civil War certainly produced its fair share of white IDPs, the plight of 4 million freedpeople is the most significant aspect of the IDP issue. Southern blacks had to determine what to do with themselves and their families when they went suddenly from indentured servants tied to a single location to some nebulous "freedom" that typically meant they must or should leave the only homes many of them had ever known.

Before the 13th Amendment was enacted, marriages between slaves were not considered legally binding and while the family unit was a bedrock of black culture, along with the church, under slavery, families lived in constant fear of separation. Herbert G. Gutman estimated in 1976 that at least 25 percent of black marriages experienced such separation.[101] During Reconstruction, however, freedpeople came before the Army and through the Freedman's Bureau and through local governments for help in locating lost relatives. This is a crucial right not to be omitted from any evaluation of the social implications of Reconstruction, as it allowed men to regain their stature as head of families, protected women from violation, and ensured children could be raised in safe and secure units, which provided the cohesion and stability needed to support new generations of black citizens. It also allowed children to settle into a homestead where they might access education.[102]

A vestige from African culture, slaves saw families in a larger context of kinship networks, which is still evident in black culture today with grandmothers and aunts and others raising their relatives with large networks living in proximity and multigenerational households.[103] These arrangements are also a result of fewer economic and educational opportunities that leave families in poverty, which is exacerbated by high institutionalized incarceration rates of black men that cause some families to develop structures beyond the nuclear household. In addition, redlining essentially ghettoized black families into small enclaves within larger cities. Freedom meant that displaced family members were now able to look for their lost relatives but typically had little knowledge of how or resources with which to do so.

Stories abound of black people traveling hundreds of miles on foot to locate lost spouses, children, parents, and siblings.[104] This was another task taken on by the Freedmen's Bureau, which received hundreds of letters from freedpeople asking for help in locating relatives. The Army was often overwhelmed by the need to protect large numbers of people traveling throughout the South to find their relatives. Many put advertisements in newspapers, even offering cash rewards for those returning lost kinsfolk, and many wrote letters to public officials asking for information and assistance to find their relatives.[105]

While few would welcome a life of slavery, without any other experience for generations, it is not likely that every newly freed person would be suddenly prepared to ask for a job at a saw mill, open up a butcher shop or decide to go to college and become a doctor or a teacher in the U.S. South of 1865. If no longer required or permitted to remain on the plantation, what was a black family to do for food, shelter and livelihood? The Southern economy was in shambles and any unskilled labor jobs available would surely go to a white man long before a black one would be considered.

While the Federal government made a strong financial commitment to infrastructure repair and expansion in the South, beyond the efforts of the Freedmen's Bureau, not enough was done in terms of **Social Reconstruction** to knit the fabric of the country back together for long-term cohesion. Despite the best efforts of those in command and the work of the Bureau, the U.S. Army's role was never clearly defined in a way that those under its care (or heel, depending upon your perspective), could relate to as a positive force for peacebuilding. It was fully unable to achieve its mission because (a) the mission was never clarified or even agreed to and (b) conflicts in northern Union politics that enabled an "unreconstructed" South to gain readmission to the Union and, ironically, to dominate its politics down to the present time. Thus, leading to today's often uttered "truism" that the North won the Civil War and lost the ensuing peace.[106]

Thus is the legacy of Reconstruction in nineteenth-century United States. Looking at our history of "reconstructing" and stabilizing other countries, most of them also victims of colonization, we see few successes with continuing ethnic and other divisive conflict, often cloaking a never-ending battle over resources. The economic struggle becomes the personal tragedy with a lack of stable governance leading to the inability for people to live in safe and secure environments.

NOTES

1. This is an ongoing theme in Gregory P. Downs, *After Appomattox: Military Occupation and the Ends of War*, 2015, Harvard University Press: Cambridge, MA.
2. U.S. Department of Defense 2016 Biennial Assessment of Stability Operations Capabilities, Office of the Assistant Secretary of Defense for Stability Operations/Low Intensity Conflict (SO/LIC), August 2017, Washington, DC, p. 7.
3. Gregory P. Downs, *After Appomattox: Military Occupation and the Ends of War*, 2015, Harvard University Press: Cambridge, MA, p. 15.
4. Gregory P. Downs, *After Appomattox: Military Occupation and the Ends of War*, 2015, Harvard University Press: Cambridge, MA, p. 8.
5. Gregory P. Downs, *After Appomattox: Military Occupation and the Ends of War*, 2015, Harvard University Press: Cambridge, MA, p. 16.
6. Mark L. Bradley, "The Army and Reconstruction, 1865–1877," part of the *U.S. Army Campaigns of the Civil War* series, Center of Military History, United States Army, 2015, pp. 33–35, available from http://www.history.army.mil/html/books/075/75-18/cmhPub_75-18.pdf. Accessed on March 10, 2017.
7. Mark L. Bradley, "The Army and Reconstruction, 1865–1877," part of the *U.S. Army Campaigns of the Civil War* series, Center of Military History, United States Army, 2015, p. 34, available from http://www.history.army.mil/html/books/075/75-18/cmhPub_75-18.pdf. Accessed on March 10, 2017.
8. Mark L. Bradley, "The Army and Reconstruction, 1865–1877," part of the *U.S. Army Campaigns of the Civil War* series, Center of Military History, United States Army, 2015, pp. 56–57, available from http://www.history.army.mil/html/books/075/75-18/cmhPub_75-18.pdf. Accessed on March 10, 2017.
9. Mark L. Bradley, "The Army and Reconstruction, 1865–1877," part of the *U.S. Army Campaigns of the Civil War* series, Center of Military History, United States Army, 2015, p. 58, available from http://www.history.army.mil/html/books/075/75-18/cmhPub_75-18.pdf. Accessed on August 6, 2019.
10. Mark L. Bradley, "The Army and Reconstruction, 1865–1877," part of the *U.S. Army Campaigns of the Civil War* series, Center of Military History, United States Army, 2015, pp. 58–59, available from http://www.history.army.mil/html/books/075/75-18/cmhPub_75-18.pdf. Accessed on March 10, 2017.
11. *Guiding Principles for Stabilization and Reconstruction*, United States Institute of Peace and United States Army Peacekeeping and Stability Operations Institute: Washington, DC, 2009, pp. 6–53, available

from https://www.usip.org/sites/default/files/guiding_principles_full.
pdf. Accessed on August 6, 2019.

12. U.S. Department of Defense 2016 Biennial Assessment of Stability Operations Capabilities, Office of the Assistant Secretary of Defense for Stability Operations/Low Intensity Conflict (SO/LIC), August 2017, Washington, DC, p. 7.

13. *Guiding Principles for Stabilization and Reconstruction*, United States Institute of Peace and United States Army Peacekeeping and Stability Operations Institute: Washington, DC, 2009, pp. 7–67–68 including endnotes 203–206, available from https://www.usip.org/sites/default/files/guiding_principles_full.pdf. Accessed on August 25, 2017.

14. The entire text can be found under the Yale Law School Avalon Project, available from http://avalon.law.yale.edu/19th_century/lieber.asp. Accessed on August 17, 2017.

15. The entire text can be found under the Yale Law School Avalon Project, available from http://avalon.law.yale.edu/19th_century/lieber.asp. Accessed on August 17, 2017.

16. Article 114 is available from https://avalon.law.yale.edu/19th_century/lieber.asp#sec4. Accessed on February 8, 2020.

17. The entire text can be found under the Yale Law School Avalon Project, available from http://avalon.law.yale.edu/19th_century/lieber.asp. Accessed on August 17, 2017.

18. The entire text can be found under the Yale Law School Avalon Project, available from http://avalon.law.yale.edu/19th_century/lieber.asp. Accessed on August 17, 2017.

19. Article 102 is available from https://avalon.law.yale.edu/19th_century/lieber.asp#sec4. Accessed on February 8, 2020.

20. "Freedmen's Bureau Acts of 1865 and 1866," United States Senate Historical Office, available from https://www.senate.gov/artandhistory/history/common/generic/FreedmensBureau.htm. Accessed on December 11, 2017.

21. Colette Rausch, *Combatting Serious Crimes in Postconflict Societies: A Handbook for Policymakers and Practitioners*, United States Institute of Peace: Washington, DC, 2006.

22. Mark L. Bradley, "The Army and Reconstruction, 1865–1877," part of the *U.S. Army Campaigns of the Civil War* series, Center of Military History, United States Army, 2015, p. 50, available from http://www.history.army.mil/html/books/075/75-18/cmhPub_75-18.pdf. Accessed on March 10, 2017.

23. U.S. Marshals Service website displaying excepts from *The Lawmen: United States Marshals and Their Deputies: 1789–1989*, by Frederick S. Calhoun accessible from https://www.usmarshals.gov/history/peril_of_your_life.htm. Accessed on June 21, 2017.

24. Author's discussion with Dr. Richard Love, U.S. Army War College Professor of Stability, on June 21, 2017 at Carlisle Barracks in Carlisle, PA.
25. Matt Matthews, "The Posse Comitatus Act and the U.S. Army: A Historical Perspective," *Global War on Terrorism Occasional Paper No. 14*, Combat Studies Institute Press, Ft. Leavenworth, KS.
26. Garrett M. Graff, "The Story Behind Bill Barr's Unmarked Federal Agents," *Politico Magazine*, June 5, 2020 provides a thorough breakdown of the available information on the numbers and unclear responsibilities of Federal troops in the nation's Capital and around the country. This is available from https://www.politico.com/news/mag azine/2020/06/05/protests-washington-dc-federal-agents-law-enforc ement-302551. Accessed on July 22, 2020.
27. Gillian Flaccus, "Federal Agents, Local Streets: A 'Red Flag' in Oregon," US News, July 21, 2020, available from https://www.usn ews.com/news/politics/articles/2020-07-20/portland-police-federal-agents-used-gas-against-protesters. Accessed on July 22, 2020.
28. Mark L. Bradley, "The Army and Reconstruction, 1865–1877," part of the *U.S. Army Campaigns of the Civil War* series, Center of Military History, United States Army, 2015, p. 16, available from http://www.history.army.mil/html/books/075/75-18/cmhPub_ 75-18.pdf. Accessed on March 10, 2017.
29. Gregory P. Downs and Scott Nesbit, "Mapping Occupation: Force, Freedom and the Army in Reconstruction," this is an entirely online, interactive database that presents data from deep archival research on troop deployments, demographics and voting patterns by month across the Southern states from April 1865 through December 1877, available from www.mappingreconstruction.org. Accessed on August 25, 2017.
30. "Whose Heritage? Public Symbols of the Confederacy," *Southern Poverty Law Center Special Report*, April 25, 2016, available from https://www.splcenter.org/20160421/whose-heritage-community-act ion-guide. Accessed on August 3, 2017.
31. Black Lives Matter official website, available from https://blacklivesma tter.com/. Accessed on July 22, 2020.
32. "Whose Heritage? Public Symbols of the Confederacy," Southern Poverty Law Center Special Report, April 25, 2016, p. 12, available from https://www.splcenter.org/20160421/whose-heritage-com munity-action-guide. Accessed on August 3, 2017.
33. Caroline Janney, *Remembering the Civil War: Reunion and the Limits of Reconciliation*, South Carolina University Press, 2013, entire book discussed the extent of the Lost Cause Myth perpetration process and its effects since Reconstruction to the present period.

34. W.E.B. Du Bois, "Of the Death of Freedom," in *The Souls of Black Folks*, 1903, republished in 2003 by Barnes & Noble Classics: New York, NY, P. 30.

35. The Tax History Project: Reconstruction to the Spanish-American War, available from http://www.taxhistory.org/www/website.nsf/Web/THM1866?OpenDocument. Accessed on December 18, 2017. The Tax History Project was established by Tax Analysts in 1995 to provide scholars, policymakers, students, the media, and citizens with information about the history of American taxation.

36. Gregory P. Downs, *After Appomattox: Military Occupation and the Ends of War*, 2015, Harvard University Press: Cambridge, MA, p. 42.

37. Calculated using www.in2013dollars.com. Accessed on December 8, 2017; Claudia Goldin and Frank Lewis, "The Economic Costs of the American Civil War: Estimates and Implications," *Journal of Economic History*, Vol. 35, No. 2 (June 1975), pp. 299–326.

38. Blight, David W. *Race and Reunion: The Civil War in American Memory*, Belknap of Harvard University Press: Cambridge, MA, 2001.

39. Roger L. Ransom and Richard Sutch, "Capitalists Without Capital: The Burden of Slavery and the Impact of Emancipation," *Agricultural History*, Vol. 62, No. 3 (Summer 1988), pp. 133–160, Table 3.

40. Robert Selph Henry, *The Story of Reconstruction*, 1999 reprint. Konecky & Konecky: New York, NY, p. 22. Originally published in 1938 by the Bobbs-Merrill Company.

41. "Carpetbaggers & Scalawags," History.com, 2014 online article, available from http://www.history.com/topics/american-civil-war/carpetbaggers-and-scalawags. Accessed on March 5, 2017.

42. Elizabeth Nix, "What's the Difference Between a Carpetbagger and a Scalawag?" Part of History Channel's *Ask History* Series, March 3, 2015, available from http://www.history.com/topics/american-civil-war/carpetbaggers-and-scalawags. Accessed on August 21, 2017.

43. "Guerilla Warfare Missouri," Civil War Trust, available from https://www.civilwar.org/learn/articles/bushwhackers-and-jayhawks. Accessed on August 21, 2017.

44. Joseph J. Thorndike, "Tax History: Can the 14th Amendment Fix Everything?" October 10, 2013, available from http://www.taxhistory.org/thp/readings.nsf/ArtWeb/D59F21D284ECDF8F85257D1B0041C86F?OpenDocument. Accessed on August 21, 2017.

45. Tax History 1866–1900: Reconstruction to the Spanish-American War, The Tax Museum, 2017, available from http://www.taxhistory.org/www/website.nsf/Web/THM1866?OpenDocument. Accessed on August 21, 2017.

46. Claudia Goldin and Frank Lewis, "The Economic Costs of the American Civil War: Estimates and Implications," *Journal of Economic History*, Vol.

35, No. 2 (June 1975), pp. 299–326; Conversion calculated using www.in2013dollars.com. Accessed on February 8, 2020.

47. Gregory P. Downs, *After Appomattox: Military Occupation and the Ends of War*, 2015, Harvard University Press: Cambridge, MA, p. 92.

48. Gregory P. Downs, *After Appomattox: Military Occupation and the Ends of War*, 2015, Harvard University Press: Cambridge, MA, p. 94.

49. Gregory P. Downs, *After Appomattox: Military Occupation and the Ends of War*, 2015, Harvard University Press: Cambridge, MA, p. 95.

50. Gregory P. Downs, *After Appomattox: Military Occupation and the Ends of War*, 2015, Harvard University Press: Cambridge, MA, p. 96.

51. Gregory P. Downs, *After Appomattox: Military Occupation and the Ends of War*, 2015, Harvard University Press: Cambridge, MA, p. 96.

52. Gregory P. Downs, *After Appomattox: Military Occupation and the Ends of War*, 2015, Harvard University Press: Cambridge, MA, p. 97.

53. "America's Reconstruction: People and Politics After the Civil War," Digital History, available from http://www.digitalhistory.uh.edu/exhibits/reconstruction/section3/section3_09.html. Accessed on August 20, 2017.

54. Mark L. Bradley, "The Army and Reconstruction, 1865–1877," part of the *U.S. Army Campaigns of the Civil War* series, Center of Military History, United States Army, 2015, p. 38, available from http://www.history.army.mil/html/books/075/75-18/cmhPub_75-18.pdf. Accessed on March 10, 2017.

55. Perkinson, Robert, *Texas Tough: The Rise of America's Prison Empire*, 2010, Metropolitan Books: New York, NY, p. 105.

56. "Report on the Investigation of the Ferguson Police Department," U.S. Department of Justice, Civil Rights Division, March 4, 2015, available from https://www.justice.gov/sites/default/files/opa/press-releases/attachments/2015/03/04/ferguson_police_department_report.pdf. Accessed on August 5, 2019.

57. Douglas A. Blackmon, *Slavery by Another Name: The Re-enslavement of Black Americans from the Civil War to World War Two*, 2012, First Anchor Books, a Division of Random House: New York, NY, 2014 Paperback edition cover graphic.

58. S. Mintz and S. McNeil. 2016. "Convict Lease System," Digital History, available from http://www.digitalhistory.uh.edu/disp_textbook.cfm?smtid=2&psid=3179. Accessed on August 9, 2017.

59. W. Fitzhugh Brundage, "Slavery by Another Name," *Created Equal, a Film Project by the National Endowment of the Humanities*, available from https://createdequal.neh.gov/community/programming-guide/programming/created-equal-scholar-essays#slavery-by-another-name. Accessed on July 22, 2020.

60. "Chain Gangs Ended in Alabama," *United Press International*, June 20, 1996, available from https://www.upi.com/Archives/1996/06/20/Chain-gangs-ended-in-Alabama/1645835243200/. Accessed on July 22, 2020.

61. James Surowiecki, "Trump Sets Private Prisons Free," *The New Yorker*, December 5, 2016, available from https://www.newyorker.com/magazine/2016/12/05/trump-sets-private-prisons-free. Accessed on September 13, 2017.

62. U.S. Department of Health & Human Services, Substance Abuse and Mental Health Services Administration, 1999 National Household Survey on Drug Abuse, August 2000, available from http://www.samhsa.gov/OAS/NHSDA/1999/Table%20of%20Contents.htm#TopOfPage. Accessed on September 13, 2017; "The Drug War Is the New Jim Crow," published in *North American Congress in Latin America (NACLA) Report on the Americas*, July/August 2001 American Civil Liberties Union (ACLU), available from https://www.aclu.org/other/drug-war-new-jim-crow. Accessed on September 13, 2017.

63. Christopher Uggen, Ryan Larson, and Sarah Shannon, "6 Million Lost Voters: State-Level Estimates of Felony Disenfranchisement, 2016," The Sentencing Project, October 6, 2016, available from http://www.sentencingproject.org/publications/6-million-lost-voters-state-level-estimates-felony-disenfranchisement-2016/. Accessed on September 13, 2017.

64. "Freedmen's Bureau," Black History Series, History.com, updated 2013, available from http://www.history.com/topics/black-history/freedmens-bureau. Accessed on December 8, 2017.

65. *Guiding Principles for Stabilization and Reconstruction*, United States Institute of Peace and United States Army Peacekeeping and Stability Operations Institute: Washington, DC, 2009, pp. 8–115, available from https://www.usip.org/sites/default/files/guiding_principles_full.pdf. Accessed on July 20, 2020.

66. "Andrew Johnson," U.S. Senate Biography, available from https://www.senate.gov/artandhistory/history/common/generic/VP_Andrew_Johnson.htm. Accessed on October 11, 2017.

67. "Freedmen's Bureau Acts of 1865 and 1866," United States Senate Historical Office, available from https://www.senate.gov/artandhistory/history/common/generic/FreedmensBureau.htm. Accessed on December 11, 2017.

68. "Andrew Johnson" U.S. Senate Biography, available from https://www.senate.gov/artandhistory/history/common/generic/VP_Andrew_Johnson.htm. Accessed on October 11, 2017.

69. "The Enforcement Act of 1871 text on Anti-Discrimination Center website," available from http://www.antibiaslaw.com/list/enforcement-act-of-1871#n2958. Accessed on December 8, 2017.

70. Mark L. Bradley, "The Army and Reconstruction, 1865–1877," part of the *U.S. Army Campaigns of the Civil War* series, Center of Military History, United States Army, 2015, p. 63, available from http://www.history.army.mil/html/books/075/75-18/cmhPub_75-18.pdf. Accessed on March 10, 2017.
71. Lisa Rein, "Civil War Gave Birth to Much of Modern Federal Government," *Washington Post*, October 7, 2011, available from https://www.washingtonpost.com/politics/civil-war-gave-birth-to-much-of-modern-Federal-government/2011/09/22/gIQA43EFSL_story.html?utm_term=.b140f46d8f0e. Accessed on August 29, 2017.
72. Pendleton Civil Service Act of 1883 full text, available from https://www.ourdocuments.gov/doc.php?flash=true&doc=48. Accessed on December 11, 2017.
73. Morrill Act signed by President Abraham Lincoln on July 2, 1862 full text, available from https://memory.loc.gov/cgi-bin/ampage?collId=llsl&fileName=012/llsl012.db&recNum=534. Accessed on August 12, 2017.
74. "1890 Land Grant History," Prairie View A&M University Library, available from http://www.pvamu.edu/library/about-the-library/history-of-the-library-at-prairie-view/1890-land-grant-history/. Accessed on December 8, 2017.
75. Howard University Founding in "Reconstruction," *America's Story*, available from http://www.americaslibrary.gov/jb/recon/jb_recon_howard_1.html. Accessed on July 22, 2020.
76. Siobhan Fitzpatrick, "The Southern Famine Relief Commission: Feeding the South While Founding Reconciliation," *Madison Historical Review*, Vol. 8, Article 1, 2011, available from http://commons.lib.jmu.edu/mhr/vol8/iss1/1?utm_source=commons.lib.jmu.edu%2Fmhr%2Fvol8%2Fiss1%2F1&utm_medium=PDF&utm_campaign=PDFCoverPages. Accessed on December 11, 2017.
77. Robert Selph Henry, *The Story of Reconstruction*, 1999 reprint. Konecky & Konecky: New York, NY, p. 422. Originally published in 1938 by the Bobbs-Merrill Company.
78. For an enormous compilation of illustrated examples, see "The U.S. Army Corps of Engineers: A History," U.S. Army Corps of Engineers: Arlington, VA, 2008.
79. Emma Sky, *The Unraveling: High Hopes and Missed Opportunities in Iraq*, 2015, Public Affairs, a member of the Perseus Books Group: New York, NY, p. 253.
80. Mark Berman, "Former Coal CEO Sentenced to a Year in Prison After 2010 West Virginia Coal Mine Disaster," *Washington Post*, April 6, 2016, available from https://www.washingtonpost.com/news/post-nation/wp/2016/04/06/former-coal-ceo-sentenced-to-a-year-in-prison-

for-2010-west-virginia-coal-mine-disaster/?utm_term=.8f2bb29c5178. Accessed on May 10, 2017.

81. Caroline Janney, "Mourning and Celebration in the Wake of War," in *Remembering the Civil War: Reunion and the Limits of Reconciliation*, University of North Carolina Press: Chapel Hill, NC, 2013, pp. 73–102.

82. Caroline Janney, *Remembering the Civil War: Reunion and the Limits of Reconciliation*, University of North Carolina Press: Chapel Hill, NC, 2013, p. 235.

83. Hanes Walton Jr., Sherman Puckett, and Donald R. Deskins Jr., *The African American Electorate: A Statistical History*, CQ Press: Thousand Oaks, CA, 2012, p. 280; Note Congress proposes Constitutional Amendments as "joint resolutions," which then must be approved or "ratified" by two thirds of the states before it can become part of the Constitution. See "Constitutional Amendment Process," National Archives Federal Register, available from https://www.archives.gov/fed eral-register/constitution. Accessed on December 8, 2017.

84. Hanes Walton Jr., Sherman Puckett, and Donald R. Deskins Jr., *The African American Electorate: A Statistical History*, CQ Press: Thousand Oaks, CA, 2012, p. 167.

85. Peter T. Dalleo, "The Growth of Delaware's Antebellum Free African American Community," U.S. Courthouse, Wilmington, Delaware, June 27, 1997, a history project of University of Delaware, available from http://www1.udel.edu/BlackHistory/antebellum.html. Accessed on August 26, 2017.

86. Gregory P. Downs and Scott Nesbit, "Mapping Occupation: Force, Freedom and the Army in Reconstruction," this is an entirely online, interactive database that presents data from deep archival research on troop deployments, demographics and voting patterns by month across the Southern states from April 1865 through December 1877, available from www.mappingreconstruction.org. Accessed on August 25, 2017.

87. Photograph taken in Charlottesville, VA by passerby Jill Mumie on a July 8 "rehearsal day" for the white supremacist rally that resulted in the death of one woman by vehicular homicide on August 13, 2017. This article is a reconstruction of the origin of the now iconic photograph of, not a police officer, but Charlottesville High School resource officer, Darius Nash. The photo is generally but erroneously believed to have been taken at the deadly event in August. This alone illustrates the difficulty in unpacking the systematic institutionalization of 300 years of persecution or one race by another that will be nearly impossible to unravel and eliminate without violent and radical upheaval, available from http://time.com/4899668/charlottesville-virginia-protest-officer-kkk-photo/. Accessed on August 15, 2017.

88. Eric Foner, *Reconstruction: America's Unfinished Revolution, 1863–1877*, Harper & Row: New York, NY, 1988, p. 143.
89. W.E.B. Du Bois, "Of the Dawn of Freedom," in *The Souls of Black Folks*, Barnes & Noble Classics: New York, NY, 2003 ed., pp. 22–23.
90. W.E.B. Du Bois, "Of the Dawn of Freedom," in *The Souls of Black Folks*, Barnes & Noble Classics: New York, NY, 2003 ed., p. 20.
91. According to www.CivilWar.org administered by the Civil War Trust, "For Every Three Soldiers Killed in Battle, Five More Died of Disease," available from https://www.civilwar.org/learn/articles/civil-war-casualties. Accessed on December 11, 2017.
92. Gettysburg, Pennsylvania certified tour guide during October 2016 walking tour taken by the author.
93. Robert F. Reilly, MD, "Medical and Surgical Care During the American Civil War, 1861–1865," Baylor University Medical Proceedings: Dallas, TX, April 2016; 29(2), pp. 138–142, available from https://www.ncbi.nlm.nih.gov/pmc/articles/PMC4790547/. Accessed on July 22, 2020.
94. *120 Years of American Education: A Statistical Portrait*, U.S. Department of Education Office of Educational Research and Improvement National Center for Education Statistics, Table 2. School enrollment of 5- to 19-year olds per 100 persons, by sex and race: 1850–1991, p. 14, available from https://nces.ed.gov/pubsearch/pubsinfo.asp?pubid=93442. Accessed on December 18, 2017.
95. Robert Selph Henry, *The Story of Reconstruction*, 1999 reprint. Konecky & Konecky: New York, NY, p. 435. Originally published in 1938 by the Bobbs-Merrill Company, p. 435: The exact 2017 equivalent was $14,125,823.98 assuming a 1.98 percent average inflation rate per year. Conversion done using www.into13dollars.com, available from http://www.in2013dollars.com/1870-dollars-in-2017?amount=785700. Accessed on August 10, 2017.
96. Robert Selph Henry, *The Story of Reconstruction*, 1999 reprint. Konecky & Konecky: New York, NY, pp. 422–436. Originally published in 1938 by the Bobbs-Merrill Company.
97. U.S. Department of Commerce, Bureau of the Census, Historical Statistics of the United States, Colonial Times to 1970; and Current Population Reports, Series P-23, Ancestry and Language in the United States: November 1979 (This table was prepared in September 1992), available from https://nces.ed.gov/naal/lit_history.asp. Accessed on July 7, 2017.
98. "The U.S. Literacy Rate Hasn't Changed in 10 Years," *Huffington Post*, December 12, 2014, available from http://www.huffingtonpost.com/2013/09/06/illiteracy-rate_n_3880355.html. Accessed on August 30, 2017.

99. Barton Myers, "Sherman's Field Order No. 15," Civil War & Reconstruction, 1861–1877 entry in the *Georgia Encyclopedia*, last updated on June 8, 2017, available from https://www.georgiaencyclopedia.org/art icles/history-archaeology/shermans-field-order-no-15. Accessed on July 24, 2020.

100. Barton Myers, "Sherman's Field Order No. 15," Civil War & Reconstruction, 1861–1877 entry in the *Georgia Encyclopedia*, last updated on June 8, 2017, available from https://www.georgiaencyclopedia.org/art icles/history-archaeology/shermans-field-order-no-15. Accessed on July 24, 2020.

101. Herbert G. Gutman, *The Black Family in Slavery and Freedom, 1750–1925*, Pantheon Books: New York, 1976, 3.

102. Eric Foner, "Building p. the Black Community: The Family," from *The Meaning of Freedom, Black and White Responses to the End of Slavery*, 2003, part of the *American Reconstruction: People and Politics After the Civil War Project* of University of Houston's Digital History Library, available from http://www.digitalhistory.uh.edu/exhibits/rec onstruction/section2/section2_family.html. Accessed on July 24, 2020.

103. Tristan L. Tolman, "The Effects of Slavery and Emancipation on African-American Families and Family History Research" *Crossroads: A Journal of African-American Research*, March 2011, available from http://www.leaveafamilylegacy.com/African_American_Fami lies.pdf. Accessed on July 24, 2020.

104. Eric Foner, "Building the Black Community: The Family," from *The Meaning of Freedom, Black and White Responses to the End of Slavery*, 2003, part of the *American Reconstruction: People and Politics After the Civil War Project* of University of Houston's Digital History Library, available from http://www.digitalhistory.uh.edu/exhibits/rec onstruction/section1/section1_family.html. Accessed on July 24, 2020.

105. Tristan L. Tolman, "The Effects of Slavery and Emancipation on African-American Families and Family History Research," *Crossroads: A Journal of African-American Research*, March 2011, available from http://www.leaveafamilylegacy.com/African_American_Fami lies.pdf. Accessed on July 24, 2020.

106. Much of this paragraph is attributed to Dr. Andy Roth, from his March 2020 review, as he said it all so much better than I could.

CHAPTER 4

Observations and Recommendations for Future Stabilization

Abstract This project began in March 2017 as the U.S. Army's Peace Keeping and Stability Operations Institute (PKSOI) considered a review and possible update to the 2009 *Guiding Principles for Stabilization and Reconstruction.* The case of post-U.S. Civil War Reconstruction was chosen, as there are longer-term outcomes to assess, rather than a currently evolving, dynamic contemporary case, such as Afghanistan or Iraq. The U.S. Army was not only the lead, but the sole stabilizing actor in concert with the Executive and Legislative Branches of the domestic Federal government following the Civil War, rather than playing a supporting role to the Department of State or the United Nations, as recommended for today's conflicts. These observations and recommendations are solely those of the author.

Keywords Discontinuity of battlespace management · Voter fatigue · Demobilization · Governance · Elite bargaining

INTRODUCTION

This project was begun in March 2017 as the Army's Peace Keeping and Stability Operations Institute (PKSOI) was considering a review and

D. E. Chido, *US Army's Effectiveness in Reconstruction
According to the Guiding Principles of Stabilization,*
https://doi.org/10.1007/978-3-030-60005-1_4

possible update to the *Guiding Principles*. This was the original intent of attempting to look at them from a perspective outside of their genesis for use in Afghanistan and Iraq. Discussion was undertaken about evaluating them in terms of a current stabilization operation, such as that in Somalia or in the Lake Chad Basin. However, it was decided that it made more sense to look at a case for which there might be less cultural complexity and longer-term outcomes to assess.

Thus, the idea to apply them to an historical period and one in which the U.S. Army was not only the lead, but the sole stabilizing actor in concert with the Executive and Legislative Branches of the stabilizing government. These observations and recommendations were the result of that review and are solely those of the author. While a great number of active and retired U.S. military officers provided insights and reviews of portions of the chapter, they are not officially endorsed by PKSOI or any other agency of the U.S. Army or Joint Force.

LACK OF PLANNING

The divisiveness of the political partisans regarding postwar policy might have been mitigated if the Army and the U.S. government had entered the occupation with a clearly articulated plan and goals for governance in both the short and long terms. However, the American Army gave no thought to the post-conflict period or its role in it until well after...surrender....[1]

This 2007 Association of the U.S. Army (AUSA) *Institute of Land Warfare Paper* quoted above is not about Afghanistan or Iraq, but about Reconstruction. This assessment could be applied to either, despite the Army producing thousands of after action reports and establishing an entire organization devoted to collecting and disseminating lessons learned since 1985.[2] DOD is also considered the planning center crucial to whole-of-government stabilization activity, yet its behavior in the lead-up to a large-scale invasion and supposed occupation of Iraq as late as 2003, a country larger than California, precisely mirrors its activity 140 years earlier with missed opportunity to learn from many additional conflicts in the meantime.

"Woefully Under-Resourced"

The Commanders of the five Southern Districts were endowed by Congress with immense powers as governors to manage law enforcement and justice administration, as well as the political process with mainly the Lieber Code as guide. As experienced in recent stabilization operations, under President Johnson's Second Reconstruction, they were only given, "20,000 soldiers to police an area roughly the size of Western Europe with a population of over 8 million."[3] While the calculations of force or police to population ratios for stabilization fluctuate, they typically average 20 (civilians) to 1 (military/police) in relatively stable areas. By 1867, this ratio was 400 to 1 and only increased through the Reconstruction period.

In the Fall and Winter of 1865, a perfect storm developed to derail the Army's positive effects postwar. The Army's mobility and manpower had been reduced by demobilization and the start of new contract negotiations for staff just as state governments began to reenter the Union, ending the Army's authority to operate in many locations. These effects led directly to increased violence in the areas where it became increasingly difficult for the Army to have a presence. Movement of troops out of the South continued this way through 1866 as troops mustered out of service and others redeployed to the West.[4]

This issue of discontinuity of battlespace management continues today with short-term deployments of units unable to form the relationships needed to be effective in their area of operations (AO). In addition, the current emphasis on lethality,[5] rather than all of the tasks the military will actually be required to perform within the current construct of the Competition Continuum in Multi-Domain Battle conditions,[6] the new conceptual model of future warfare, will again leave commanders and soldiers woefully under-resourced for effectively doing their jobs.

Voter Fatigue

The problem of voter fatigue with stabilization did not first emerge in Vietnam. It has been part of American politics at least since the Civil War and historians would likely agree that it is eternal. Only 10 years after the end of this very bloody domestic affair, the criticality of a long-term commitment to ensure sustainable peace was not recognized. When asked for Federal troops to end a massacre in the rural area near Clinton, Mississippi, President Grant responded, "The annual Autumnal outbreaks and calls for troops are getting to be nauseating to the American people."[7] It must be emphasized that these "outbreaks" and "calls" were all *by* American people, not from some unknown foreign would-be Host Nation seeking assistance in quelling rebellion, where the American military more clearly has the right to refuse.

The issue for today's *stabilizers* is that U.S. voters became disinterested in these activities intended to ensure a stable society for all of *its own* citizens in the wake of a brutal war at home, it seems decision makers should not expect the progeny of these same voters to maintain support for foreign interventions once conflict has ended for the same time period or for the decades that have characterized successful post-conflict stabilization activities as in Germany and South Korea.

The U.S. Civil War did not last just four years from 1861 to 1865 but roughly 16 years, from 1861 until 1877 when U.S. Army troops withdrew. Just as the Iraq war, which began in March 2003, did not end when President George W. Bush infamously declared less than two months later, "Major combat operations in Iraq have ended" aboard the USS *Abraham Lincoln* before a White House provided banner proclaiming "Mission Accomplished."[8]

The official end came on December 31, 2011 when the "last American soldiers cross[ed] the border out of Iraq" as President Barack Obama announced on December 12, 2011 from the White House in a joint press release with Iraqi Prime Minister Nouri al-Maliki, "After nearly nine years, our war in Iraq ends this month." President Obama continued by saying, "Millions have cast their ballots – some risking or giving their lives – to vote in free elections. The Prime Minister leads Iraq's most inclusive government yet. Iraqis are working to build institutions that are efficient and independent and transparent."[9]

This process begun by the withdrawal of Union troops is directly akin to the way Iraqi "terror" began to rise almost immediately as the United States withdrew its troops and was solidified by ISIS enacting widespread violence across the region and taking control of the Iraqi city of Mosul and eventually other areas from the Iraqi Army and declaring these territories part of its "Islamic Caliphate" less than three years later in June 2014.[10] Thus, the most common failure of stabilization operations is bringing it to an end before sustainable peace has been achieved.

As is evident from Afghanistan and Iraq, despite Inspector General (IG) reports of ineffectual activities and waste, fraud and abuse, the consensus is building to identify the large amount of time it takes to achieve stability after such devastation, when the violence may have largely ended, but the divisions and grievances that caused it to remain, festering into poor governance, continuing disenfranchisement of some groups and poor economic recovery for significant portions of the population.

When President Trump declared in December 2017 that ISIS had been defeated and that U.S. combat operations in Syria would now come to an end, Brett McGurk, U.S. special envoy to coalition forces said, "The United States is prepared to remain in Syria until we are certain that ISIS is defeated, stabilization efforts can be sustained, and there is meaningful progress in the Geneva-based political process." The more experienced Secretary of Defense, recently retired James Mattis had earlier explained, that U.S. forces would "stay in Syria as long as Islamic State fighters want to fight and prevent the return of an 'ISIS 2.0,'" meaning it was now time for stabilization activities and the DDRR process to begin.

After the president's proclamation, Mattis softened on using the military for this effort, saying, "What we will be doing is shifting from what I would call an offensive, shifting from an offensive terrain-seizing approach to a stabilizing … you'll see more U.S. diplomats on the ground." In the same press conference, he clarified, "Well, when you bring in more diplomats, they're working that initial restoration of services…The military would move our diplomats around, make certain they're protected. You know, they've got bodyguards around them, and there's still bad guys out there…The longer term recovery is going to take a lot of effort and a lot of years."[11]

The *2016 DoD Biennial Assessment* refers to the "stabilization stigma," as a misunderstanding of the need for stabilization activities before, during, and after conflict, not just after conflict has ended to ensure national security objectives are met. The *Assessment* coherently describes the precise roles of the U.S. military and other actors including other U.S. government agencies that may be better equipped for long-term engagement, such as USAID, as well as expectations of host nations, who should be involved in security cooperation activities and joint training long before a crisis, and thus be prepared to take the lead for stabilization at home with the United States only providing support. This "stigma" prevents effective policy development in the pre-conflict phase and highlights how prescient President Lincoln was in planning for Reconstruction long before the meeting at Appomattox.

The Civil-Military Dilemma
and the Structure of Governance

General Sherman, for one, believed that the Army's approach to the population should be clearly dichotomous in cases of war and not war. As Sherman said of Army activities from 1863 through 1865, "During the war the military is superior to civil authority, and where interests clash the civil must give way." This included activities in which commanders "directly emancipated slaves, seized food, exiled traitors, shuttered newspapers, hanged outlaws, starved regions, and other efforts to 'sap Confederates' will or capacity to fight.'" Of the post-conflict period, however, Sherman opined, "Where there is no conflict, every encouragement should be given to well-disposed and peaceful inhabitants."[12]

While reviews of the efficacy of Reconstruction are mixed and multitudinous, those that focus at all on the Army's role tend today to see it as critical. James Sefton wrote that the Army was "by far the most important instrument of Federal authority in the South" and that "it was the only enforcer of national reconstruction policy."[13] The Army's ability to ensure civil rights for freedpeople depended upon its accessibility to them for prevention of atrocities, redress of outrages and enforcement of laws protecting their rights and property.

As the Army was redistributed throughout the Reconstruction period, its reach into areas where freedpeople were concentrated was uneven. As cavalry troops were gradually moved to Texas and further West to deal with Native and Mexican incursions and crimes, such as horse thievery, freedpeople's ability to access the Army was reduced. This is illustrated in Fig. 4.1, produced by the author from Gregory Downs' massive dataset on Reconstruction-era statistics, which visually indicate, among a great many other things, the disposition of Army installations over time.

As Gregory Downs noted in the introduction to his statistics, "U.S. power existed where the government could enforce its laws through the Army. [Its] installations formed the centers of patchwork zones of occupation. These zones were linked and extended by the Southern rail network as the Army responded to political pressure and events on the ground. Mapping also visualizes the more limited areas from which black Southerners could reach soldiers, highlighting the unequal geography of the Army and the civilians who used it to assert their rights."[14]

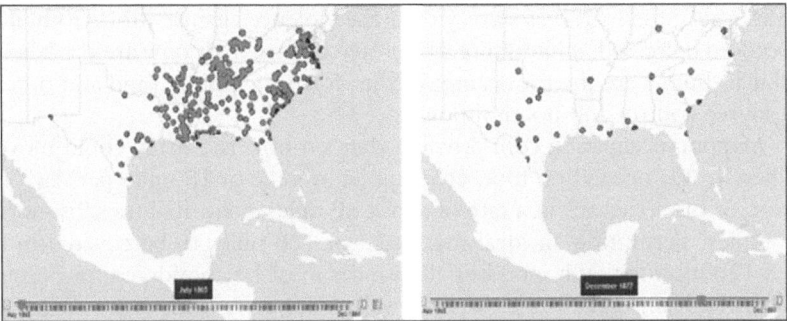

Fig. 4.1 Comparison of zones of army access in May 1865 and December 1877

The Army not only read aloud the Emancipation Proclamation throughout the South, it also served to instill "practical freedom." Former slaves knew and helped Army officers understand that

> As planters whipped ex-slaves, barred them from leaving, refused them pay, and tortured them, ex-slaves ran to Army posts and convinced officers there that to be meaningful, freedom would have to include more than the right not to be sold. In posts scattered across places like western Georgia, or South Carolina, the Army remained a point of access for freedpeople into the Fall. There, Army officers, sometimes working through the Freedmen's Bureau, took control of court cases involving freedpeople, arrested white vigilantes, and tried to stop the resurgent practices of slavery.[15]

GOVERNANCE

The terms military *government* or *governance* are often used interchangeably with military *support to government*. There is little understanding outside the practitioner circle to accurately describe the institutional development, ministerial advising, personnel training, and other governance activities ongoing in Afghanistan, Iraq, and many other places today. U.S. policy has followed the Reconstruction model more closely than, perhaps, an historical view would recommend. In the contemporary cases the United States strove to support the central governments of the two countries invaded in 2001 and 2003 respectively and to help them eradicate radical sectarianism and Islamic terrorism. In the post-Civil

War South, the issue was slavery or as the violence against efforts to build freedom increased, white supremacist "terrorism." Not only are the labels similar, but so are the root causes of the violence: identity, land and other resource control and political power.

Mappingoccupation.com provides data on how the Army could move where it was needed on foot, covering an average of 18 miles per day on foot, or on horseback at a rate of about 30 miles a day. Rail travel is more complex, permitting an ideal maximum of 120 miles to be covered in a day (20 mph for six hours) but allowances must be considered for points of dis-and em-barkation as well as wait times at stations and vast areas where tracks and trains did not reach or were in disrepair. This resulted in a more realistic (but perhaps still optimistic) estimated average of 5–10 miles per day the Army could cover to maintain its influence.

In contrast, as freedpeople were less likely to have horses and were unlikely to attempt to use the railroads, it is assumed that they could likewise walk 18 miles a day to reach the Army and lodge complaints or provide warnings. Downs estimates the average speed by any conveyance was closer to about six miles per hour.[16] This difficulty in covering territory increasingly contributed to the ability of mounted violent white supremacists to attack and disappear with impunity as the number of overall and particularly cavalry troops dropped. The areas where the Army was unable to penetrate with consistency quickly developed various alternative governance structures, depending upon the demographic and geographic disposition of each.

The issue of "ungoverned spaces," where the "legitimate" Federal or even state authority does not reach is not widely understood. Human social organization does not permit power vacuums to exist for long and wherever the Union Army reach was sparse or nonexistent in this period, alternatives to its governance capability invariably arose.

The use of the Army as the "long arm of the law" reaching across the South was highly effective in ensuring the Federal government could provide security and the population could access its services, however, its reach could not be fully ubiquitous as resources were constantly constrained.[17] The Army's increasing difficulty in covering territory contributed to the ability of mounted violent white supremacists to attack and disappear with impunity as the number of overall, and particularly, cavalry troops was reduced.

Geography plays a crucial role in the government's ability to prevent the rise of alternative governance. Not only is distance a challenge, but the ruggedness of the terrain can prohibit movement and divide populations within a single region into distinct and often competing sub-units. Mountainous, desert, and jungle terrain particularly have this effect. Looking at U.S. geography, the areas around the Appalachian Mountains have remained relatively fractious throughout U.S. history, as did those in the Southwestern desert and Rocky Mountain areas as settlers moved west and encountered diverse Mexican and Native groups with varying interests in sharing the goals of Federal governance. Such features closely resemble those in familiar areas of Afghanistan and Iraq.

Minority Shi'a and black disenfranchisement were respectively embedded in Iraq's and the American South's very institutions and founding. When these groups were empowered both by Washington's intervention and violent overthrow of the regimes that maintained both Sunni and white elite status, the Army was needed to establish new constitutions and institutions to safeguard these new rights and to ensure the old guard relinquished its grip. Both of these processes would take vastly more time and cost more than the American public or the Army were willing to contribute and as William Blair noted, "the central role of the military has been underappreciated in the histories of Reconstruction, [which] rarely give the Army its due as the central agent for social and political change."[18]

In the contemporary landscape, examples of alternative governance abound, including where the Taliban, rather than the Afghan central government, controls security, trade, employment, education, healthcare, and justice at all levels. In Colombia, the Revolutionary Armed Forces of Colombia (FARC) and various drug cartels controlled large swathes of territory for more than half a century. As the 2016 peace treaty with the FARC comes into force, DDRR programs strive to alleviate concerns that former members who have no skills other than providing security to outlaws will move to work with new organized crime syndicates that will take over the drug trade or move to nearby countries where they can continue to ply their trade as mercenaries, thus bringing conflict back to the country.

Under these conditions, the government of Peru, for one, is endeavoring to extend Federal reach and reduce alternative governance by transnational organized criminal (TOC) groups through its "Mobile State" program. This program deploys riverboats to remote jungle areas to bring federal services ranging from healthcare to social services to providing access to voter registration and identification cards to mobile cash machines.[19]

HOME RULE OF LAW

The U.S. Marshal model might be one to consider for stabilization post-conflict or even for stabilizing fragile states. In the United States, from their founding in 1789, the Marshals served their local area providing federally mandated and legalized police service, living within their jurisdictions and often being originally from these areas. This gives them local situational understanding of the players, their grievances, aspirations, and influences, as well as knowledge about their perceptions of state and national issues, making them naturally able to anticipate unrest. Marshals were also permitted to hire their own deputies and did not report to a central headquarters until 1969, further establishing their autonomy within a national construct.

When situations called for it, as in the 1969 Vietnam War protests at the Pentagon, a thin row of Marshals blocked the entrance with a robust line of Army soldiers behind them. Only the Marshals have the power to arrest, but the Army can back them up if they request assistance, essentially possessing the capability to provide transitional public security until the local or state civilian authorities can maintain full control.

The Marshals believe that "their mission is national; their methods are local." On the other hand, in contentious periods or divided communities, the local aspect of the Marshal dispensation can be a complicating factor. For instance, numerous Marshals across the South resigned just before the start of the Civil War, choosing their local communities over the Federal system they had sworn to serve. This localized and autonomous system can also be subject to patronage if the hiring process is too decentralized and if their mandate is not clear that, while serving as a member of the community, their primary role is to enforce the national legal code.[20]

EMBRACE STABILIZATION AS A MISSION

Some have suggested that the attitude of the Army commanders tasked with stabilization has an effect on the success of their mission, as DiMarco argued, "One reason for this failure was that the Army did not embrace the mission. Army commanders, even those who proved dedicated and talented administrators of civil government in the South, hated the mission."[21] This is yet another comment that might be easily mistaken for a recent assessment of activities during the Global War on Terror.

DiMarco posits that local conditions as well as domestic conditions in the occupying nation itself can conspire to negatively affect the trajectory of stabilization, again discussing Reconstruction, but sounding like the assessment is related to post-9/11 activities when he stated that "national partisan politics, local political opposition, lack of planning, terrorism, lack of resources and lack of will all contributed to the Army's inability to achieve and sustain success...."[22]

Conditions in the occupying power's own body politic are extremely likely contributors to the failure of Reconstruction and stabilization writ large. Partisan politics lead to a lack of clear and unified objectives for the stabilization process itself, which results in poor planning and a lack of will to allocate the necessary types and amounts of resources and eventually ends in a fruitless effort and a failure to "consolidate gains," which some policymakers and the public can observe and say, "See, we don't do nation building and this is why."

As General Sherman said of post-U.S. Civil War Reconstruction, "No matter what change we may desire in the feelings and thoughts of the people [in the] South, we cannot accomplish it by force."[23] This is what makes the issue of "Political Primacy" so complicated.

The Army was the instrument of choice to advise in Korea and Vietnam, but the State Department led the postwar efforts to stabilize Germany and Japan, only using the military as needed for security and logistics. The Army is still in the lead in Afghanistan and Iraq, with yet another troop surge sent in August 2017 to "finish" the 16-year old "job" in Afghanistan. As President Trump announced this surge of 4000 troops with the express job of "killing terrorists" and not "nation building," it is not clear what, if any, elements of the *Strategic Framework* will ultimately be employed in a post-conflict Afghanistan to effectively "win" the peace.

GUIDING PRINCIPLES: PROPOSED REVISION AND APPLICATION

Recommendation #3 from the "Whole of Government" section of the *2016 Biennial Assessment* suggests that DoD "Work with the U.S. Institute of Peace (USIP) and the 3-D [Diplomacy-Development-Defense]

Stabilization Working Group to review, update and document the *Guiding Principles for Stabilization and Reconstruction.*"[24]

As described above, this monograph applied the *Strategic Framework* as if it were an analytic framework, which it is not intended to be. One criticism of the *Guiding Principles* is that their constituent end states are not measurable. Even if they are applied to a varying degree to each case, there should be an understood level of each that can be measured. How much is "stable enough" governance, for instance?

As defined in the *Guiding Principles*, "End states represent the ultimate goals of a society emerging from conflict." This definition alone is nebulous as an entire society is not monolithic. The majority of the society may want stability, but a vocal or violent segment of the population will ensure this is never achieved if its own critical objectives are not met. This aim can be in direct opposition to the other end states desired by the rest of society, such as independence for the group and territory in question.

In the U.S. Civil War, slavery was the central, intractable issue, an economic necessity for the South that spawned other elements of conflict, such as racist oppression, lack of will to diversify economic development, cultural and identity politics and violent extremism, which only increased with postwar occupation. The requirement for the Federal government to conduct stabilization operations led to enduring resentment that prevented full integration of freedpeople into civic life and maintained a suspicion of government among whites, due to occupation and among blacks, due to the later institutionalization of anti-civil rights efforts of the Southern Democrats, both of which are still pervasive today.

Below the level of end states, the *Framework* provides conditions, which should be measurable and consist of "approaches," which are essentially tasks that can be assessed as completed or not completed, depending upon the expectations of condition level, which may or may not vary for each stabilization case (see *Annex I: Approaches, Conditions and End states for Stabilization and Reconstruction from the Guiding Principles*). As noted by Aditi Gorur of the Stimson Center, when assessing the use of the term in UN mandates, "stabilization...is not a discrete task but a strategic objective"[25] which, here, are comprised of its constituent end states and their constituent approaches, which can be viewed as tasks.

Framework for Developing Stabilization Task List and Evaluation Tool

Of course, a much closer assessment of a specific end state or condition within an end state by a specialized expert in governance or economics, or whichever end state requires scrutiny, as well as expert knowledge about the stabilization case would have a far more detailed and nuanced perspective on what went "right" and "wrong." A more scientific and detailed approach to evaluation would also yield far more significant and actionable insights, but this is simply a proposed starting point. This model suggests a way to reimagine the *Principles* as a tool for historical research, case comparison, and assessment to determine the effectiveness of past, current, or even future interventions.

In addition to developing a qualitative assessment metric, this approach can also enable the *Guiding Principles* to evolve beyond a basic manual into a tasking and priorities checklist. This could further evolve into a stability measurement tool, akin to the Fund for Peace's *Fragile States Index*.[26] This would give those involved in stabilization efforts the capacity to show measurable progress to those who provide the authorities and resources to enable continuation of the process through the very long-term commitment required for success at any measure. It could also help policymakers determine which cases are worth intervening in and whether a military-led security approach is needed or whether the destabilizing issue is economic with a more appropriate USAID or Treasury role to shore up a fragile state or prevent recurrence of violence post-conflict.

The first step to developing this *Framework* into these sorts of tools would be to **standardize the language** used to express the approaches to the end states. For instance, under Rule of Law—Public Order, what does "a comprehensive system" mean? How could this be measured? A system for what precisely?

Second, the third-tier **approaches should lead with a verb** to indicate an action that, once completed, has contributed to achievement of the second-tier condition. This results in a clear list of tasks to be accomplished in order to systematically achieve the condition they are designed to support.

Third, the measurement process for effective achievement of each condition should clearly indicate whether **commitment to the given process is expected to be long- or short-term in duration**. For instance, "Separate warring parties" as an approach to the "Cessation of large-scale

violence" under the "Safe and Secure Environment" end state is an immediate short-term task that would be a first priority, required for almost
any other activity to take place. It would become a long-term task if a
permanent cordon was needed because the key drivers of conflict were
not effectively addressed and the violence or threat of violence continued,
ultimately thwarting efforts at stabilization, as in the contemporary case
of Mali, which the UN calls a "stabilization mission," as there is no peace
treaty, and thus no parameters for peace to be kept.[27]

On the other end of the time continuum would be "Achieve monetary
stability," which could take several decades, thus delaying the attainment
of "Macroeconomic stabilization" as a condition required for a Sustainable Economy. This is why observers often point to post-World War II
Germany and Japan as stabilization cases that "went right," implying that
we have forgotten how to "do stabilization," but the reality is that a
decades-long commitment was made to these cases that was even absent
in the U.S.' own post-Civil War Reconstruction obligation.

Finally, it is important that the **appropriate joint, interagency, inter-
governmental, and multinational (JIIM) partners** are engaged for
activities to which they are best suited. The U.S. Army should not be
responsible for the long-term economic process suggested above when
the International Monetary Fund (IMF) and World Bank exist precisely
to provide this type of assistance and both organizations anticipate such
development efforts to require very long-term commitment, as does
USAID. Long-term goals and gains in stabilization can often be subverted
by short-term actions intended to deal with critical current circumstances
without consideration of longer-term effects of such actions.

General Sherman noted that "hearts and minds cannot be changed
through force."[28] This is what makes the issue of "Political Primacy" so
complicated. The Army can maintain order but cannot perform all the
activities needed to ensure sustainable stability, so what exactly is its role?
Mark Bradley summed it up perfectly

> Almost from the start, the Army in the South lacked the force to fill that
> role adequately. By the 1870s, peacekeeping invariably meant protecting
> blacks and white Republicans from the night riders and the rifle clubs.
> While the results were uneven at best, the Army was all that stood
> between the freedpeople and the Democratic terrorist organizations. The
> withdrawal of federal troops in 1877 thus ensured the triumph of white
> supremacy in the South. Nevertheless, the use of the U.S. Army to enforce

the Reconstruction Acts had established a precedent that would resonate eight decades later, when President Dwight D. Eisenhower would deploy units of the 101st Airborne Division to enforce the Federally mandated desegregation of Central High School in Little Rock, Arkansas.[29]

The Army was also the instrument of choice a decade and a half earlier to stabilize postwar Germany and Japan, and then Korea and then Vietnam. The Army is probably inappropriately still in the lead in Afghanistan and Iraq, with yet another troop surge announced in August 2017 to "finish" the 16-year old "job" in Afghanistan.

Address Elite Bargaining as Part of Political Primacy

Although Stable Governance advocates "Political Moderation and Accountability," a recent study on elite bargaining from the United Kingdom Stabilization Unit (UKSU) looked at 15 case studies of post-conflict resolution and had identified the importance of not excluding existing elites from the bargaining table, even when it seems they are at fault for the violence. They recommend current stabilization doctrine, like the *Guiding Principles*, to focus on establishing peace agreements and enhancing formal government institutions, but UKSU has found that in some post-conflict societies, it is not through these formal institutional systems, such as constitutional reform and electoral democracy where the battles for power are waged.

Thus, efforts to "tame" elites to prevent them from interfering with interventionist peace efforts, are not likely to be as effective as recognizing that if they are satisfied with other benefits, resource allocations, and opportunities, such as in business or politics, they are likely to participate more effectively in the new systems. When they are "excluded or their interests are ignored, the prospects for stabilizing violent conflict and sustaining peace diminish."[30]

This perfectly mirrors the situation during Reconstruction, when powerful Southern elites, politicians and slave-owning landholders worked through political violence to reduce the power blacks and unaligned whites could wield through the ballot box and in positions of authority to reduce the effects of Reconstruction, while they waited it out for their anticipated Redemption period, when they could return to complete political, economic, and social domination in their region and beyond.

CONCLUSION

Thus, the U.S. Army worked tirelessly to stabilize the U.S. South after the Civil War. It did not have the resources nor the capability to conduct the required tasks, but through ingenuity and contracting, as well as volunteers willing to do their part, the Army did the best it could in the short-term while Reconstruction was underway. Once the mandate to remain was lifted, the Redemption period returned with Southern Democrats regaining power and freedmen losing it for another century. Due to the failure to fully stabilize the South and integrate black citizens into the life of the country, we still have enormous disparities in income, opportunity, incarceration, education, health outcomes, and social cohesion with rising white supremacy and violence today.

Stability has been essentially maintained as the South and the Union as a whole have institutionalized prejudices that prevented black citizens from full participation until the second Civil Rights law was passed in 1964, but the system and structure of the political, economic, and judicial system retains its inherent biases, resulting in recurrent social upheaval and violent clashes as in the 1960s and 1970s with the Black Panther and other black revolutionary groups and again seeming to attempt to reassert these rights with the thus far nonviolent Black Lives Matter movement, which is primarily directed at police and criminal justice reform. Along with economic shifts, these have contributed to a resurgence in white supremacist activity as well, which is likely to further threaten stability far into the twenty-first century unless U.S. institutional changes are made today to alter these structural, historical, unreconstructed conditions.[31]

NOTES

1. Louis DiMarco, "Anatomy of a Failed Occupation: The U.S. Army in the Former Confederate States, 1865 to 1877," *Land Warfare Paper No. 66W*, November 2007, The Institute of Land Warfare Association of the United States Army (AUSA), p. 5, available from https://www.ausa.org/publications/anatomy-failed-occupation-us-army-former-confederate-states-1865-1877. Accessed on August 26, 2017.

2. Center for Army Lessons Learned (CALL) is part of the Combined Arms Center (CAC) located at Ft. Leavenworth in Kansas, accessible from https://www.ausa.org/publications/anatomy-failed-occupation-us-army-former-confederate-states-1865-1877. Accessed on July 11, 2017.

3. Mark L. Bradley, "The Army and Reconstruction, 1865–1877," part of the *U.S. Army Campaigns of the Civil War* series, Center of Military History, United States Army, 2015, p. 8, available from http://www.his tory.army.mil/html/books/075/75-18/cmhPub_75-18.pdf. Accessed on March 10, 2017.

4. Gregory P. Downs and Scott Nesbit, "Mapping Occupation: Force, Freedom and the Army in Reconstruction," this is an entirely online, interactive database that presents data from deep archival research on troop deployments, demographics and voting patterns by month across the Southern states from April 1865 through December 1877, available from www.mappingreconstruction.org. Accessed on August 25, 2017.

5. U.S. National Defense Strategy, 2018, U.S. Department of Defense, available from https://www.tradoc.army.mil/Portals/14/Documents/MDO/TP525-3-1_30Nov2018.pdf. Accessed on August 6, 2019.

6. The U.S. Army in Multi-Domain Operations 2028, TRADOC Pamphlet 525-3-1, December 6, 2018, available from https://www.tradoc.army.mil/Portals/14/Documents/MDO/TP525-3-1_30Nov2018.pdf. Accessed on August 6, 2019.

7. Mark L. Bradley, "The Army and Reconstruction, 1865–1877," part of the *U.S. Army Campaigns of the Civil War* series, Center of Military History, United States Army, 2015, p. 69, available from http://www.his tory.army.mil/html/books/075/75-18/cmhPub_75-18.pdf. Accessed on March 10, 2017.

8. "White House Pressed on 'Mission Accomplished' Sign: Navy Suggested It, White House Made It, Both Sides Say," Dana Bach, CNN.com, October 29, 2003, available from https://www.cnn.com/2003/ALL POLITICS/10/28/mission.accomplished/. Accessed on February 9, 2020.

9. "Remarks by President Obama and Prime Minister al-Maliki of Iraq in a Joint Press Conference," December 12, 2011, available from https://oba mawhitehouse.archives.gov/the-press-office/2011/12/12/remarks-pre sident-obama-and-prime-minister-al-maliki-iraq-joint-press-co. Accessed on February 9, 2019.

10. "Timeline: The Rise, Spread, and Fall of the Islamic State," *Insights and Analysis*, The Wilson Center, October 28, 2019, available from https://www.wilsoncenter.org/article/timeline-the-rise-spread-and-fall-the-islamic-state. Accessed on February 9, 2020.

11. *Department of Defense Press Gaggle* with Secretary Mattis, December 29, 2017 transcript, available from https://www.defense.gov/Newsroom/Transcripts/Transcript/Article/1406278/press-gaggle-with-secretary-mattis/. Accessed on July 22, 2020.

12. Gregory P. Downs, *After Appomattox: Military Occupation and the Ends of War*, 2015, Harvard University Press: Cambridge, MA, pp. 22–23.

13. James E. Sefton, *The United States Army and Reconstruction, 1865–1877*, Louisiana State University Press: Baton Rouge, 1967, p. ix.
14. Gregory P. Downs and Scott Nesbit, "Mapping Occupation: Force, Freedom and the Army in Reconstruction," this is an entirely online, interactive database that presents data from deep archival research on troop deployments, demographics and voting patterns by month across the Southern states from April 1865 through December 1877, available from www.mappingreconstruction.org. Accessed on August 25, 2017.
15. Gregory P. Downs and Scott Nesbit, "Mapping Occupation: Force, Freedom and the Army in Reconstruction," this is an entirely online, interactive database that presents data from deep archival research on troop deployments, demographics and voting patterns by month across the Southern states from April 1865 through December 1877, available from www.mappingreconstruction.org. Accessed on August 25, 2017.
16. In order to calculate an average distance Union soldiers could cover in a day, Gregory Downs and his team used archival evidence to make some basic assumptions on mobility and the Army's ability to protect vulnerable populations and redress atrocities. "We assumed that soldiers could march three miles per hour, or that they could commandeer a railroad car and after two hours of transition time, ride it at 20 miles per hour, with a total travel time in all transportation modes allowed of six hours per day. Where cavalry were stationed, we assumed they could cover 5 miles in an hour. We do not possess data on all rail depots for the South, and so we made the assumption that soldiers could embark and disembark at any point along a rail line. These assumptions are not ironclad—some soldiers clearly could travel farther, some not nearly so far, and sites of embarkation were likely not nearly so flexible. But they allow us to map a broad region of partial control, where the army might be able to exert itself if called upon," available from http://mappingoccupation.org/index.html (Narrative section, page 5). Accessed on December 11, 2017.
17. See MappingOccupation.org for an exhaustive study and interactive timelines and visualizations on Civil War Reconstruction data to include Army mobility, available from www.mappingoccupation.org. Accessed on December 11, 2017.
18. "The Use of Military Force to Protect the Gains of Reconstruction," William Alan Blair, *Civil War History*, Vol. 51, No. 4, December 2005, p. 390.
19. "New Developments in Organized Crime in Peru," R. Even Ellis, *Cipher Brief*, May 20, 2016, available from https://www.thecipherbrief.com/column_article/new-developments-in-organized-crime-in-peru. Accessed on February 9, 2020.
20. U.S. Marshal Service website displaying excerpt from *The Lawmen: United States Marshals and Their Deputies: 1789–1989*, by Frederick S.

Calhoun, available from https://www.usmarshals.gov/history/loyal_com munity.htm. Accessed on June 21, 2017.

21. Louis DiMarco, "Anatomy of a Failed Occupation: The U.S. Army in the Former Confederate States, 1865 to 1877," *Land Warfare Paper No. 66W*, November 2007, The Institute of Land Warfare Association of the United States Army (AUSA), p. 3, available from https://www.ausa.org/publications/anatomy-failed-occupation-us-army-former-confederate-states-1865-1877. Accessed on August 26, 2017.

22. Louis DiMarco, "Anatomy of a Failed Occupation: The U.S. Army in the Former Confederate States, 1865 to 1877," *Land Warfare Paper No. 66W*, November 2007, The Institute of Land Warfare Association of the United States Army (AUSA), p. 3, available from https://www.ausa.org/publications/anatomy-failed-occupation-us-army-former-confederate-states-1865-1877. Accessed on August 26, 2017.

23. Mark L. Bradley, "The Army and Reconstruction, 1865–1877," part of the *U.S. Army Campaigns of the Civil War* series, Center of Military History, United States Army, 2015, p. 72, available from http://www.his tory.army.mil/html/books/075/75-18/cmhPub_75-18.pdf. Accessed on March 10, 2017.

24. *U.S. Department of Defense 2016 Biennial Assessment of Stability Operations Capabilities*, Office of the Assistant Secretary of Defense for Stability Operations/Low Intensity Conflict (SO/LIC), August 2017, Washington, DC, p. 11.

25. Aditi Gorur, "Defining the Boundaries of UN Stabilization Missions," Stimson Center, December 2016, p. 23, available from https://www.stimson.org/content/defining-boundaries-un-stabilization-missions. Accessed on August 26, 2017.

26. Fund for Peace *Fragile States Index*, available from https://fragilestatesin dex.org/. Accessed on July 22, 2020.

27. United Nations Multidimensional Integrated Stabilization Mission in Mali (MINUSMA) Fact Sheet, available from https://peacekeeping.un.org/en/mission/minusma. Accessed on July 22, 2020.

28. Mark L. Bradley, "The Army and Reconstruction, 1865–1877," part of the *U.S. Army Campaigns of the Civil War* series, Center of Military History, United States Army, 2015, p. 72, available from http://www.his tory.army.mil/html/books/075/75-18/cmhPub_75-18.pdf. Accessed on March 10, 2017.

29. Mark L. Bradley, "The Army and Reconstruction, 1865–1877," part of the *U.S. Army Campaigns of the Civil War* series, Center of Military History, United States Army, 2015, p. 72, available from http://www.his

tory.army.mil/html/books/075/75-18/cmhPub_75-18.pdf. Accessed on March 10, 2017.

30. Christine Cheng, Jonathan Goodhand, and Patrick Meehan, *Synthesis Paper: Securing and Sustaining Elite Bargains That Reduce Violent Conflict*, UK Stabilisation Unit, April 2018, available from https:// assets.publishing.service.gov.uk/government/uploads/system/uploads/ attachment_data/file/765882/Elite_Bargains_and_Political_Deals_Pro ject_-_Synthesis_Paper.pdf. Accessed on July 20, 2020.

31. For more on this, see "How to Distinguish Between Antifa, White Supremacists, and Black Lives Matter," Conor Friedersdorf, *The Atlantic*, August 31, 2017, available from https://www.theatlantic.com/politics/ archive/2017/08/drawing-distinctions-antifa-the-alt-right-and-black-lives-matter/538320/. Accessed on December 8, 2017.

Annex I: Approaches, Conditions, and End States for Stabilization and Reconstruction from the Guiding Principles

Cross-Cutting Principles
Conditions for Achieving
HN Ownership and Capacity
Understanding local context
Fostering local ownership
Inclusivity
Capacity building
Formal & informal systems
Early resources
Role of women
Effective transitions from international to host nation
Political Primacy
Using a conflict lens
Fostering and sustaining a political process
Inclusivity of warring parties and marginalized groups
Effective strategic communications
Legitimacy
A bargain between citizens and the government
Mandate and authorities
Matching resources to goals and delivering a timely peace dividend
Peace process leadership
Accountability and transparency
Management of expectations and communication

(continued)

© The Editor(s) (if applicable) and The Author(s), under exclusive license to Springer Nature Switzerland AG 2021
D. E. Chido, *US Army's Effectiveness in Reconstruction According to the Guiding Principles of Stabilization*,
https://doi.org/10.1007/978-3-030-60005-1

(continued)

Cross-Cutting Principles

Constituencies for peace
Engagement of the international community
Unity of Effort
A shared understanding of the situation
A shared strategic goal
Integration
Cooperation and coherence
Civil-military cooperation
Recognition of humanitarian space
Security
Information
Management of spoilers
Reform of the security sector
Protection of human rights
Conflict Transformation
Understanding drivers and mitigators of conflict
Reducing drivers of conflict and strengthening mitigators
Building HN capacity to manage conflict drivers via nonviolence and long-term development support
Regional Engagement
Conduct diplomacy to obtain support for stable HN & region
A shared regional vision
Cooperation

Detailed Guidance Assessment: Safe and Secure Environment

Conditions for Achieving Endstates	Guidance to Approach
Cessation of Large-Scale Violence	
Separation of Warring Parties	Move quickly to separate warring parties and stop the violence
	Separate forces to create time and space for the peace process
	Apply principles of restraint, impartiality, and consent when dealing with parties to the conflict
Enduring Cease-fire/Peace Agreement	Understand that stopping armed conflict requires political, not military solutions
	Transform the conflict to pursuit of political and economic goals by nonviolent means
Management of Spoilers	Anticipate obstructionists and understand their motivations
	Create a plan for managing spoilers through inclusion or exclusion as appropriate
	Maintain the primacy of the peace process
	Adopt an "assertive position" with regard to peace agreement enforcement
Intelligence	Remember the population is the best information source
	Be aware of local sensitivities
	Given the sensitivities, be creative in acquiring critical information
	Coordinate military-police intelligence sharing
	Develop capacity to conduct intelligence-led operations against spoilers through Disruption, Dislocation or Decisive action
Legitimate State Monopoly over Means of Violence	
Disarmament and Demobilization	Start DD planning early
	Tailor DD strategy to local conditions
	Include details of disarmament and demobilization in the peace agreement
	Provide credible security guarantees to build confidence in disarmament
	Maximize host nation ownership in the disarmament and demobilization strategy

(continued)

(continued)

Detailed Guidance Assessment: Safe and Secure Environment

	Inform the population to build popular support
	Aim for inclusivity of all warring parties
	Include affected nontraditional combatants
	Ensure accountability to human rights standards through identification
	Ensure that DD is civilian-led, with technical input and operational support from international forces
Reintegration of Ex-Combatants	Prepare for reintegration to be the most sensitive and difficult phase of DDR
	Avoid making ex-combatants a privileged class by integrating them into broader recovery strategies aimed at all conflict-affected populations
	Sustain international support for the reintegration process
Security Sector Reform	Ensure that reforms reflect the security needs of the host nation population
	Strengthening security forces is not enough; promote good governance and legitimatecivilian oversight to ensure long-term accountability.
	Prevent infiltration of security forces through robust vetting
	Focus on public service ethos and competence when training security forces
	Support the improvement of police-community relations and police responsiveness
	Ensure coherence of strategy and effort among major actors
	Promote the civil authority of the state; long-term stability depends on it
Physical Security	
Security of Vulnerable Populations	Respect the boundaries of humanitarian space and understand humanitarian principles of independence, humanity, impartiality, and neutrality
	Ensure that the UN mandate includes the obligation to protect civilians under imminent threat of physical violence
	Protect vulnerable public officials
	Address all aspects of landmines

(continued)

(continued)

Detailed Guidance Assessment: Safe and Secure Environment

Protection of Infrastructure	Coordinate across military, law enforcement, and civilian actors to provide security
	Protect and promote safety of cultural and historical sites to mitigate conflict
	Protect high-value infrastructure targets to prevent disruption to peace
Protection of War Crimes Evidence	Prioritize witness protection programs to ensure willingness of people to testify
	Move quickly to secure crime sites to avoid tampering or sabotage by spoilers
Territorial Security	
Freedom of Movement	Facilitate movement for people and goods
	Deny movement to opponents of the peace
	Be aware of cultural sensitivities when conducting checkpoint, cordon and search,
	and convoy operations
Border Security	Pay attention to border issues; they are oft overlooked at the peril of the mission
	Address border security in the mandate, cease-fire, and peace agreements
	Be prepared to perform border security functions for an indeterminate period.
	Build host nation capacity for border security as a first-order priority
	Use existing models for regional cooperative trade programs
	Manage border relations with neighbors

Detailed Guidance Assessment: Rule of Law

Conditions for Achieving Endstates	Guidance to Approach
Just Legal Frameworks	
Legal Framework Assessment	Gather, catalogue, and distribute the applicable laws first
	Conduct a comprehensive analysis of the applicable law
	Conduct a comprehensive analysis of the applicable law
	Realize the inherent constraints of new laws if they are not enforced
Short-Term Law Reform	Consider the need for legal restatement in the aftermath of conflict
	Undertake discreet legal reform in the short-term if necessary
Law Reform Process	Support and engage in a transparent and participatory process
	Decide upon a reform strategy and establish a coordinating body
	Set realistic time frames
	Use outside experts wisely
	Engage multidisciplinary, multi-skilled teams
	Conduct an impact assessment of new draft laws, and factor training into the strategy
	Ensure a sound promulgation and publicization process
Content of New Laws	Think about the details of drafting early on in the process
	Use international standards as the normative framework for law reform
	Forget the common law and civil law debate, and think about hybridization
	Consider how to appropriately use foreign laws to inform the process
	Consider the relationship between the formal and informal justice sectors when determining new content
Public Order	
A Comprehensive System	Take a holistic approach when developing a strategy for public order
	Transform systemic threats to public order as a prerequisite
Interim Law Enforcement	Move quickly to prevent criminal elements and political spoilers from cementing their grip on power
	Be prepared to perform critical law enforcement functions when necessary

(continued)

(continued)

Detailed Guidance Assessment: Rule of Law

	Coordinate public order functions between the military and the police and ensure that any gap is filled
	Deploy local forces whenever possible without compromising human rights and justice
	Address illegal armed groups and informal policing structures
	Support the building of host nation police capacity
Interim Judiciary	Deploy core elements for an emergency judiciary
	Employ informal systems when it makes sense
	Protect the judiciary from physical harm and outside influence
	Ensure that witnesses and victims are adequately protected and supported
	Vet the judiciary
	Use mobile courts and paralegals to meet immediate needs
Humane Detention and Imprisonment	Meet minimum international standards for treatment of detainees
	Take measures to provide for prison security
	Address illegal and excessive pretrial detention and prison overcrowding
	Improve prison conditions
	Prevent torture and focus on evidence-based criminal investigation
Accountability to the Law	
Transitional Justice	Protect and preserve evidence
	Choose the most appropriate transitional justice options
Horizontal and Vertical Accountability	Promote the separation of powers and judicial independence
	Support the use of horizontal accountability mechanisms
	Provide for vertical accountability mechanisms
	Consider international engagement as a necessary safeguard for accountability
Access to Justice	
Equal Access	Address barriers to both quantity and quality

(continued)

(continued)

Detailed Guidance Assessment: Rule of Law

	Enhance physical access
	Increase access through provision of legal aid
	Promote legal awareness
	Strengthen civil society as the foundation for promoting access to justice
	Recognize that increased access to justice depends on public confidence in the justice system
Remedies for Grievances	Understand informal justice mechanisms
	Use the local context to determine how and to what extent local practices should be incorporated into the formal legal system
	Modify or use informal systems in combination with formal mechanisms to ensure adherence to international human rights standards while maximizing access and public trust in the system.
	Support the adjudication of claims for a remedy through the formal state justice system and civil society
	Support the adjudication of claims for a remedy through the informal non-state justice system
	Develop culturally acceptable alternatives to harmful practices
	Support the enforcement of remedies
Fairness	Ensure equal application of the law
	Promote procedural fairness
	Facilitate transparency in all judicial processes
	Ensure effective application of the law, ensure adequate authority to enforce judgments, and improve the efficiency of court administration and management
	Increase the knowledge and professionalization of justice personnel to dispense justice

Culture of Lawfulness

Participation and Communication	Support legal empowerment of marginalized communities

(continued)

(continued)

Detailed Guidance Assessment: Rule of Law

Education and Culture	Promote public participation
	Promote communication between the justice system and the population
	Ensure transparency
	Support school-based education
	Involve centers of moral authority
	Engage the mass media and popular culture
	Work with law enforcement agencies

Detailed Guidance Assessment: Stable Governance

	Guidance to Approach
Conditions for Achieving Endstates *Provision of Essential Services* **Core Service Delivery**	Focus on providing security, the rule of law, economic governance, and basic human needs services for stability and to provide space for political settlements and development. Transparency and accountability mechanisms help ensure that the government delivers essential services effectively and reliably. Understand the roles of state and non-state actors in providing services and the impact of those actors on stability Deliver security as a top priority and provide the cornerstone for stable governance Rebuild and uphold the rule of law as a primary responsibility of the host nation government. Provide good economic governance as a framework for stabilization and reconstruction. Deliver essential services to meet basic human needs and restore the basis for government legitimacy
Access and Non-Discrimination	How essential services are provided is just as important as the delivery itself Provide equal access to services and nondiscrimination in delivery to enhance the government's legitimacy, support the peace process, and help prevent a renewal of conflict
Host Nation Capacity	Build host nation capacity to deliver essential services in a professional, accountable, and sustainable manner Make peace pay through effective personnel management Manage expectations of the population through communication about service delivery.
Stewardship of State Resources	

(continued)

(continued)

Detailed Guidance Assessment: Stable Governance

Restoration of Executive Institutions and Public Administration	Understand the terrain
	Prepare for transitional governance, but keep a focus on permanent governance
	Ensure local participation within transitional governance structures through consultation or co-administration
	Restore managerial capacity for governance
	Reform national ministries and public administration to ensure accountable use of public resources and use of regulatory power in a nondiscriminatory manner
	Focus on civil servants
	Develop the top-down and bottom-up political processes and institutional structures that are required for stable governance
	Strengthen subnational governance capacity
	Consider the impact of different forms of decentralization on stabilization
Security Sector Reform	Prioritize good governance of the security sector
	Establish accountable civilian authority over the security sector to protect human rights and prevent the renewal of conflict
	Strengthen legislative, judicial and civil society participation and oversight to prevent abuse of power
	Ensure that the host nation population drives governance reform of the security sector, as it is an inherently political process
Protection of State Resources	Promote good economic governance to enable recovery and generate confidence in the government's ability to manage public finances

(continued)

(continued)

Detailed Guidance Assessment: Stable Governance

Address low-level corruption that deprives the government of badly needed resources

Sever the nexus between government officials and illicit sources of revenue

Establish oversight mechanisms in government processes to ensure accountability

Make government financial data and activities as clear and open as possible for the population

Keep budget deficits under control by mobilizing revenue and increasing the efficiency of the tax system

Protect natural resources as fundamental state assets that are integral to economic recovery and political stability

Political Moderation and Accountability
National Constituting Processes

Build broad-based consensus on the country's political future through inclusive and participatory national constituting processes

Help generate agreement on central issues for governance to prevent the renewal of violent conflict

Focus on the process for writing the constitution as much as what the constitution says

Political Governance and Conflict Management

Help former warring factions to reframe their interests through non-violent political processes

Bring the widest range of leaders into the political process and seek to include voices of moderation

Reinforce issue-based politics over identity politics

Systems of Representation

Meet requirements for free and fair elections in order to reflect the population's interests

(continued)

(continued)

Detailed Guidance Assessment: Stable Governance

Legislative Strengthening

Consider the timing and impact of elections on the stability of the host nation

Consider the design and structure of the legislature to aid in stabilization

Strengthen legislative bodies to counterbalance the executive branch and help bolster representative and accountable governance

Train and mentor legislators and staff for conflict management

Civic Participation and Empowerment
Civil Society Development

Leverage existing capacities in developing civil society

Establish a legal and regulatory framework to protect CSOs and ensure they are allowed to form and operate freely

Foster ownership of host nation CSOs by providing necessary support to boost capacity

Promote inclusivity in developing CSOs

Foster and support community-based development to broaden civic participation and enhance opportunities for developing leadership in civil society

Promote accountability of CSOs through regulatory oversight mechanisms

Independent Media and Access to Information

Nurture and sustain a media sector that is pluralistic, transparent, sustainable, and independent

Consider creating media monitoring mechanisms to prevent incendiary or hate speech from destabilizing the country

Ensure that media outlets are representative of and accessible to the population

(continued)

(continued)

Detailed Guidance Assessment: Stable Governance

	Define media broadly but distinguish carefully between media sector development and strategic communications
	Develop a strong legal framework to protect the rights of journalists
	Encourage the development of journalism training and education programs to promote journalistic standards and potential for long-term success
	Complement education programs by creating professional associations for journalists to connect host nation actors with the international media network
Inclusive and Participatory Political Parties	In developing political parties, foster inclusivity but prioritize the commitment to peace
	Pay special attention to engaging women, minority ethnic groups, and other marginalized populations in the development of political parties
	Provide political parties with necessary training and support, but ensure neutrality in delivering that support

Detailed Guidance Assessment: Sustainable Economy

Conditions for Achieving Endstates	Guidance to Approach
Macroeconomic Stabilization	
Monetary Stability	Assess the state of monetary stability
	Inform monetary decisions by setting up a system for collecting economic data
	Address macroeconomic stabilization early on; it is an oft-overlooked priority
	Build public confidence by stabilizing domestic currency
	Stabilize the exchange rate through a foreign exchange market
	Set realistic targets for inflation rates
	Build the institutional capacity of an independent and credible monetary authority
	Strive for relevance, transparency, and effectiveness when developing a banking system
Fiscal Management	The fiscal authority should be effective and transparent
	Do not ignore revenue generation strategies to meet urgent needs in these environments.
	Stress simplicity in developing tax systems and policies, given limited administrative capacity
	Accept low tax rates on earned income in the emergency phase
	Consider debt relief programs to achieve debt sustainability
	Be prepared to fill the gap between revenue and government costs with funding from international actors
	Strengthen public expenditure management (PEM) of the host nation government
	Prioritize transparency in contracting and procurement practices to combat corruption

(continued)

(continued)

Detailed Guidance Assessment: Sustainable Economy

Legislative and Regulatory Framework

Reflect national interests in budget and state spending

Assess legal conditions and simplify wherever possible

Promote predictability, open markets, and fair competition through commercial laws

Develop a customs, tax, and budget legal framework to govern fiscal operations

Prioritize dispute resolution mechanisms to address property and contract issues

Focus on laws to combat organized crime and other illicit economic activity

Engage the private sector on advocacy for policies and regulatory reform

Control Over the Illicit Economy and Economic-Based Threats to Peace

Control Over Illicit Economic Activity

Understand the different economic channels that emerge from violent conflict

Understand the legacy of the war economy and its effects on stabilization and reconstruction

Prioritize the identification and disruption of finance networks of local power brokers, insurgent groups, transnational organized crime, and terrorist organizations

Consider the consequences of aggressively curbing the informal economy

Understand the consequences of predatory local actors in managing economic recovery programs

Recognize that the public sector can be a major source of corruption

Deal with the harmful use of remittances

(continued)

(continued)

Detailed Guidance Assessment: Sustainable Economy

Management of Natural Resource Wealth

Understand the context before designing a strategy to manage natural resources

Prevent control over natural resources, resource-rich areas, and relevant facilities by predatory actors

Draw from past approaches for improving the management of resource wealth and cutting off financing for hostile actors

Strengthen governance practices to improve natural resource wealth management

Maximize participation from all players to ensure effective management

Reintegration of Ex-Combatants

Assess the social and economic situation to identify the best options for reintegrating ex-combatants

Be sure to address the needs of communities receiving ex-combatants

Market Economy Sustainability
Infrastructure Development

Establish infrastructure priorities according to broader strategic imperatives

Consider social, political, and environmental impacts in designing infrastructure projects

Prioritize power, roads, ports, and telecommunications

Protect infrastructure to ensure peace and economic growth.

Focus on infrastructure management, not just infrastructure itself

Rebuild only what should be rebuilt

Recognize the wider benefits of infrastructure repair beyond its physical value

Private Sector Development

Look to local investment and resources; don't wait for foreign investment

(continued)

(continued)

Detailed Guidance Assessment: Sustainable Economy

Human Capital Development	Provide access to immediate credit and financial services for micro, small, and medium enterprises
	Create an enabling environment that lowers risks, promotes business activity, and attracts FDI
	Leverage key markets as economic opportunities
	Maximize the peacebuilding benefits of private sector activity
	Give all a stake in the peace process, including the most vulnerable
	Establish a means for collecting labor market data as the basis for human capital development
	Do not neglect the need to impart "life skills"
	Begin training and education on the job and look to long-term educational capacity
Financial Sector Development	Be prepared for a banking system that is severely debilitated
	Seek to rebuild trust in the banking system
	Understand that microfinance can contribute to growth, but alone it is not a substitute for reconstituting the core banking capacity of the country
Employment Generation	
Quick Impact	Generate positive results by focusing on public works projects
	Keep sustainability in mind, but avoid placing undue emphasis on it in the very early stages of recovery
	Recognize the potential impact of the international presence on economic distortions
Agricultural Rehabilitation	Provide broad assistance in rehabilitating the agricultural sector
	Avoid disincentivizing local farming through relief operations

(continued)

(continued)

Detailed Guidance Assessment: Sustainable Economy

Livelihood Development

Recruit capable, accountable individuals for a lean and effective civil service

Focus on agriculture, construction and service sectors, which will often provide the bulk of job opportunities

Pay special attention to women in micro-enterprise or vocational training

Detailed Guidance Assessment: Social Well-Being

Conditions for Achieving Endstates *Access to and Delivery of Basic Needs Services* **Appropriate and Quality Assistance**	**Guidance to Approach**
	Provide assistance based on the needs of conflict-affected populations to ensure equal access for all
	Tailor assistance to local culture
	Discourage the population from using coping strategies that arise from the inability to access basic services
	Do no harm
	Prioritize immediate relief, but do not neglect the impact on long-term development
	Coordinate humanitarian assistance and development strategies to maximize coherence and sustainability
Minimum Standards for Water, Food, and Shelter	In the emergency phases of recovery, strive to meet the immediate survival needs of the population for water, food, and shelter
	Provide quantity and quality of water to ensure survival, improve hygiene, and reduce health risks
	Impart important information to the public about the benefits of water and sanitation services and facilities
	Tailor water and food distribution and assistance according to local factors
	Use food assistance strategies that facilitate sustainability
	Aim for equity in food and water distribution
	Resort to providing free food aid only when the need is severe and there is no other alternative
	Develop tailored sanitation programs to best benefit the population
	Provide shelter assistance to meet survival needs
	When choosing a site for mass shelter, pay close attention to land rights
	Tailor shelter designs and planning to local requirements

(continued)

(continued)

Detailed Guidance Assessment: Social Well-Being

Minimum Standards for Health Services

Use shelter construction processes as an opportunity to build host nation capacity and promote livelihood development

In addition to housing, be prepared to provide nonfood items that may be necessary to maximize self-sufficiency and self-management

Treat those with the most immediate health risks while restoring basic health services for the broader population

Support a sustainable health care system for the population

Work closely with host nation health authorities and affected populations to ensure that critical needs are met

Respond appropriately and adequately to victims of sexual and gender-based violence

Restore information systems to promote public health

Access to and Delivery of Education

System-Wide Development and Reform

Use a "community-based participatory approach

Assess the context-specific relationship between education and conflict

Develop both a short-term plan for emergency action and a long-term plan for education reconstruction and development

Insulate the education system from politics

Reduce systemic corruption in the education system

Equal Access

Ensure equal access as a mitigator of conflict

Provide interim emergency education for children

Incorporate higher and nonformal education

Pay attention to refugees and IDPs

View education as a tool for child protection and welfare

Construct appropriate educational facilities

Quality and Conflict-Sensitive Education

Develop appropriate resource standards and monitor resource use

Ensure that curricula promote peace and long-term development

(continued)

(continued)

Detailed Guidance Assessment: Social Well-Being

Return or Resettlement of Refugees and IDPs

Safe and Voluntary Return or Resettlement

Enrich curricula with education on life skills

Develop and support quality teachers and administrators

Promote a student-centered, participatory learning environment

Understand the situation on the ground in order to plan effectively

Ensure voluntary return for refugees and IDPs

Ensure safety of return for refugees and IDPs

Provide refugees and IDPs with full access to the information they need to decide whether or not to return

Develop internal resettlement alternatives for those who decide not to return to their original homes

Manage refugee returns as far from the border as possible

Property Dispute Resolution

Address property disputes to encourage the return of displaced populations

Base resolution processes in a legal framework to ensure consistency and enforceability

Return property lost during conflict to its original owners where possible and offer compensation for those who must resettle

Ensure property rights of women, orphans, and other vulnerable populations

Allocate properties for community and commercial uses as needed

Reintegration and Rehabilitation

Promote self-reliance and empowerment of refugees and IDPs to prevent dependency on aid

Recognize that displaced populations represent a rich body of potential human and material assets and resources

Create an environment that sustains return

Through conflict-sensitive development, strive to build the following characteristics in the returning or resettling populations

(continued)

(continued)

Detailed Guidance Assessment: Social Well-Being

Social Reconstruction	
Inter- and Intra-Group Reconciliation	Assess existing sources of conflict to restore social capital and promote reconciliation
	Understand the cultural context to shape strategies for promoting reconciliation
	Build on indigenous practices for healing and acknowledging wrongdoing
	Ensure host nation ownership over the reconciliation process
	Recognize that reconciliation is an ongoing process—not an end goal—that may last for generations
	Pay attention to sequencing
	Consider the many different strategies that exist to promote reconciliation processes
	Be prepared to provide necessary security
Community-Based Development	Build relationships and trust through collaborative development processes
	Understand that the development process is as important as the projects
	Provide resources to ensure sustainability
	Ensure inclusion and transparency to promote reconciliation and healing

INDEX

© The Editor(s) (if applicable) and The Author(s), under exclusive   121
license to Springer Nature Switzerland AG 2021
D. E. Chido, *US Army's Effectiveness in Reconstruction*
According to the Guiding Principles of Stabilization,
https://doi.org/10.1007/978-3-030-60005-1